AF290390

INTELLIGENCE

OR CHAOS

The Atheist Delusion

The scientific proof and rational conclusion that intelligence and not chaos is the driving force behind the universe.

Henk J. Keilman

Intelligence or Chaos?

Copyright © 2022 Henk J. Keilman

Author: Henk J. Keilman

First Edition in English

No part of this publication may be reproduced, in any form or by any means – electronic, mechanical, photocopying, recording or otherwise – without the prior written permission of the publisher.

ISBN 9780952749288

Published by Ahimsa Publications.
info@motilalbooks.com

Available in ebook format ISBN 9780952749271

Cover Design & Layout: Raivata Das

TABLE OF CONTENTS

CHAPTER 1

The mystery of existence — 1

CHAPTER 2

Intelligence or chaos – the teleological argument — 21

CHAPTER 3

Materialism—matter is the only reality 63

CHAPTER 4

Darwin's theory of evolution 85

CHAPTER 5

The defects of materialism   119

CHAPTER 6

The origin of existence has to be 'simple' — 193

CHAPTER 7

The imperfect universe — 217

CHAPTER 10

The ultimate complexity – consciousness   291

CHAPTER 13
The Nature of God — 363

CHAPTER 14
Our destination: Vaikuntha — 385

I am greatly indebted to A.C. Bhaktivedanta Swami, my spiritual master and guide, who inserted real meaning into this complex journey we call life, and who shed light on this complex and mysterious universe we live in. He has been the main source of inspiration to write this book. I am also indebted to members of the Bhaktivedanta Institute, an institute founded by Srila Prahhupada, aimed at achieving a convergence of science, philosophy and spirituality. In particular I am indebted to Mr. Richard Thompson (Sadaputa Das), who was a brilliant scientist and mathematician, an eminent philosopher of science and dedicated to the mission of the Bhaktivedanta Institute. Already in the early eighties Mr. Thompson authored several books on information theory in relation to complex structures, such as "Mechanistic and Non-Mechanistic Science" a pre-cursor to the then newly emerging Intelligent Design movement. I similarly want to extend my deep appreciation to Mr. Michael Cremo, (Drutakarma das), a brilliant scientist and author of several scholarly books, including the book "Forbidden Archaeology", which he co-authored with Mr. Thompson.

Over the years while working on this book many others have

contributed, either in ideas, feed-back, criticisms or supportive arguments. I thank them all. In particular I want to thank my son Baladeva Keilman for assisting in the English editing, as well as Mr. Alister Taylor, (Advaita Chandra das), for many years of friendship, philosophical and spiritual inspiration, and for his assistance with the publishing of this book.

ABOUT THE AUTHOR

Henk Keilman was born on October 7, 1953 in Hoofddorp, a small town near Amsterdam, the Netherlands. After his high school education, he became fascinated by philosophy in general and Indian philosophy in particular. He grew up during the turbulent sixties of the previous century, a time in which the post-war generation rebelled against the established norms and values in search for alternative values. While some became involved in movements that were seeking political change, he became enchanted by the spirituality and philosophy of ancient India. In 1971 he met his spiritual master A.C. Bhaktivedanta Swami, and became initiated into the Vaishnava Vedanta tradition, one of the oldest schools within the complex religion of Hinduism. He made several journeys to India in order to further study this tradition, and to study the various philosophical schools that were part of India's rich cultural, philosophical and spiritual heritage. He was particularly impressed by the Vedanta school, its intellectual depth and its clear overlap and connection with modern science. In the past 40 years he has continued his studies and research into western and Indian philosophy and into modern science, particularly physics and cosmology. The achievements of modern science, the insights and knowledge, as well as the technological ingenuity they have produced, are absolutely impressive and historically unparalleled. What particularly impressed him was the direct link of Vedanta philosophy with the most advanced insights into physics and cosmology. He was also intrigued by the rise of atheism and its apparent support by science. Richard Dawkins bestseller "the God Delusion" was an additional trigger to present a different narrative, laying bare many of the unscientific and illogical assumptions propounded by Dawkins. Intelligence or Chaos is also a rebuke, entirely based on scientific and common sense arguments, of Dawkins his crude and unvarnished atheism. Of western thinkers Henk was inspired by the British philosopher John Locke, the great physicist Albert Einstein, and nuclear physicist Fritjof Capra. Among eastern philosophers Ramanuja and Caitanya were great sources of knowledge and inspiration. He started writing Intelligence or Chaos in 2010, the culmination of many years of research and study.

In daily life Henk Keilman is an entrepreneur and investor. In 2009 he published a book about the economic crises of 2008, called "Crisis, What Crisis", in which he compared the 2008 crises with the Great Depression of the 1930's and the lessons that can be learned from both crises. While he has a great interest in economy, technology and all the social developments related thereto, he is particularly interested in the historical and strategic intellectual forces, that have shaped and that make up the core values and norms of our present society.

Intelligence or Chaos?, A review by Michael A. Cremo

There has always been more than one way to look at the world, more than one cosmology. Some of these cosmologies emphasize matter and its random, chaotic transformations and some of them emphasize consciousness and its intelligent transformations of matter. Hence the title of this book, Intelligence or Chaos.

In any society one of these cosmologies may become dominant in terms of its cultural influence, with the alternative cosmologies held by marginalized people. Today, all around the world, a materialistic cosmology has become dominant among elites in government, economics, education, science, and the arts. Of course this has had an effect on the general population. Elements of more spiritual cosmologies remain, here and there. But the materialistic cosmology is dominant. It is characterized by the claim that all features of material reality, including consciousness, can be explained by matter operating on its own, without guidance. Basically, it's chaos, although scientists might be able to perceive some regularities that have arisen. This cosmology is compatible with atheism, and has been explicitly adopted by atheists. From this has developed a culture that is extremely materialistic.

Many have found life in a materialistic culture stifling, unsatisfying, and restrictive. Even if they were born in nominally religious families, few elements of the cosmologies of those religions have any place in public life. They are not satisfied by atheism or a weak deism, in which God remains as simply an observer of an unplanned and unguided cosmos. Such people feel there must be another way, and thus many have turned to the remaining elements of authentic spiritual cosmologies that continue to exist in various parts of the world or have explored the deeper more spiritual roots of their own traditions.

But many such people find themselves in a challenging situation. Although they have committed themselves to a cosmology that tells us that the universe is not produced by chaos, that there is a higher benevolent intelligence that plans and guides it, they remain surrounded by a world very much under the influence of the materialistic cosmology. Thus, to maintain their own commitments, and perhaps also to influence others who might be feeling restricted by the dominant materialistic cosmology, such people may find it desirable to consider rational arguments in favor of intelligence over chaos.

And that is what the author of this book has given us. He is of European heritage, from the Netherlands. He was educated according to the standards of the dominant materialistic cosmology, but eventually became dissatisfied with it and turned for guidance to a representative of one of the spiritual cosmologies that remain present on earth, the Vedic cosmology. He became a disciple of His Divine Grace A. C. Bhaktivedanta Swami Prabhupada, the founder-acharya of the International Society for Krishna Consciousness. In his teachings, Prabhupada emphasized that the Vedic cosmology is rational. It can be presented in terms of science and philosophy, as the author of Intelligence or Chaos does, accepting also the truths of other theistic traditions than his own.

The author's treatment of the paradoxical nature of Vedic cosmology, which involves not a strict dichotomy between conscious intelligence and unconscious matter, but their simultaneous oneness and difference, is thought provoking.

Also, his treatment of theodicy, the problem of evil in this world, which for many people poses the biggest obstacle to accepting the existence of God, is excellent.

In Intelligence or Chaos, the author deals with foundational issues in science, philosophy and theology in a lucid manner that is relevant not only to those following his particular Vedic path but to anyone on any theistic path. I could imagine that agnostics or even some atheists might find it of interest.

Michael A. Cremo (Drutakarma Dasa)
Los Angeles, September 4, 2020

INTRODUCTION

Several sources have inspired me to write this book. First of all from an early age I had a general interest in science and philosophy. I was fascinated by science and technology and was driven by a quest for knowledge. I was particularly interested in the philosophical implications of scientific theories. Einstein's vision of the universe for instance, intrigued me greatly. The fourth dimension of time-space, the mother of all equations $E=MC^2$, the constant speed of light, the mystery of atomic power, the possibility of being able to travel in time; all these new scientific discoveries made the universe a lot more interesting than the boring mechanical laws of Newton. Because of Einstein the universe acquired a kind of magical aura. The discoveries in the field of astronomy were equally fascinating. The inconceivable, immeasurable size of space, the gigantic galaxies, the discovery of the Big Bang, the mystery of dark matter and dark energy, reinforced the enigmatic nature of the universe. But when I tried to visualize these discoveries and put them into a meaningful perspective, I ran into some serious obstacles. Especially my attempt to visualize the laws of relativity, for instance, were problematic. Einstein's worldview was abstract, mathematical, and was in many ways beyond the capacity of visualization. While I realised I was not the only one encountering this problem that was of little consolation. My inability to develop a meaningful conceptual image of reality, and my inability to create a comprehensive philosophical understanding of the universe, frustrated my quest for knowledge.

I was not interested in science or looking for knowledge just for the sake of knowledge. I was looking for knowledge in search for a deeper and comprehensive understanding of reality. Wonderful, that four dimensional reality and impressive that sub atomic world. But what does it mean to me, what is the meaning of it all and why does it exist in the first place. Despite all the technological developments and new revolutionary scientific discoveries, the meaning of it all remained beyond my reach. In fact, as the quantity of scientific discoveries increased, and new theories were introduced, a comprehensive understanding of the universe and of life appeared ever more elusive. I also noticed that 20th and 21st century thinking was permeated with scepticism and nihilism. You can't blame scientists for being sceptical. Scientists generally tend to limit themselves to their specific disciplines, without philosophical panoramas. But even the philosophers refrained from expressing any real vision and viewpoints, and had surrendered themselves to the men and women in the white laboratory coats. Philosophy had become purely relativistic and nihilistic, from the existentialism of Jean Paul Sartre, the logical positivism of Carnap, to the language philosophy of Wittgenstein. They all echoed the same thing, there are no answers, only questions and even more questions. The famous 20th century philosopher Bertrand Russell wrote in his book "The problems of philosophy" the following on this subject: "to summarize our discussion on the value of philosophy: philosophy is not be studied for the sake of finding definitive answers to it's questions, because, as a rule, we can never know definitive answers to be true or not."

This scepticism frustrated me. What was the use of science or philosophy if they do not provide a deeper and comprehensive insight into life and existence. As I mentioned before, for me the search for science and knowledge, was primarily a search for the meaning of life. In this search I have gone through many phases. When I was fourteen I became an atheist, especially as a reaction to the, in my opinion, dogmatic teachings of Christianity. This 'conversion' was short-lived, I have never been able to fully embrace atheism. Intuitively I felt that chance and chaos could never produce what I considered to be 'my life' with all of its attributes, such as family, friends, school, my town, this planet and the rest of this vast universe. When I was fifteen I became acquainted with eastern philosophies. It was the time the Beatles were meditating in India, and the western world experienced an influx of gurus and teachers from India. Indian culture was hot. I found India to be more than a superficial and fashionable trend, and was particularly impressed with the Vedanta philosophy and the Yoga teachings, that form the bases of Hinduism. What profoundly impressed me was its intellectual depth, and the practical methods to develop knowledge and advance consciousness. Although the latter turned out to be a bit more difficult than what I expected or had hoped for.

This leads me to my second source of inspiration, which is my acquaintance with the Vaishnava Vedanta philosophy in 1970 and 1971, and my initiation into this tradition by A.C. Bhaktivedanta Swami Prabhupada. The depth, wisdom and intelligence of this worldview have inspired me profoundly. The unconditional commitment to goodness and devotion, and the emphasis on the compatibility of reason and spirituality were a revelation. For this insight and inspiration I am greatly and eternally indebted to Srila Prabhupada. In one of his famous statements Srila Prabhuada proclaimed that "Religion without philosophy leads to fanaticism, sectarianism and sentimentalism, whereas philosophy without religion will lead to endless mental speculation." Vaishnava Vedanta philosophy was able to put scientific achievements into a greater and meaningful perspective, without actually diminishing its contributions or criticizing the scientific method. To the contrary, I was very impressed with the compatibility with modern science. The Vedanta philosophy, for instance, acknowledged three sources of legitimate knowledge:

sense perception (pratyaksa in Sanskrit), logic (anumana) and the mystic experience (sabda). By the explicit acknowledgement of sense experience and logic as valid sources of knowledge, the Vedanta philosophy has avoided the mistake that has been made by many other systems of religious thought. That is the mistake of rejecting scientific knowledge, knowledge based on empirical sense perception and logic. The rejection of empirical knowledge and logic is the most important characteristic of fundamentalism. This is often accompanied by a rigid, literal interpretation of certain texts or scriptures. Fundamentalist interpretations of an ideology are not limited to religious worldviews. Any ideology that rejects or subordinates reason entirely to such texts or scriptures, that subordinates human happiness to a 'higher goal', and whose means are justified by its ends, is fundamentalist. Nazism and Marxism are good examples of atheistic ideologies that are inherently fundamentalist, or that can only be interpreted in a fundamentalist manner. 'Mein Kampf' or 'Das Kapital' are as sacred for respectively Nazis and Marxists as the Bible or Koran are for Christians and Muslims.

As I studied various scientific disciplines more deeply, I was struck by the fact that modern scientific discoveries were most compatible with the deeper meanings of Vedanta philosophy. The New Physics based on quantum theory, Einstein's relativity theory and the Big Bang cosmology, point to a non-materialistic, energetic universe. The New Physics reveal a universe that is fundamentally mysterious, and permeated by unlimited levels of complexity. Logic, mathematics and common sense dictate that this complexity must be the result of an unlimited intelligence, operating in the background. Intelligence as the driving force behind the universe is also the essence of the Vedanta philosophy. According to this philosophy the universe is ruled by a Being endowed with infinite consciousness and an inconceivable intelligence. We ourselves are a tiny part of this infinite field of consciousness. The relationship between this infinite Being and the tiny part of this Infinite Being, is one of the cornerstones of the Vedanta philosophy. This relationship cannot be captured in normal logical principles, because it is fundamentally paradoxical. The paradox is that the tiny part is simultaneously one and different from the whole. According to the normal laws of logic an object is either one or can be divided in parts. It cannot be both one and divisible. But that is exactly the case, when discussing the relationship between the part and the whole, between man and God. We are simultaneously one with and different from God, because both part and whole possess the most distinguishing of all attributes within the universe, which is consciousness. Consciousness inherently implies a degree of freedom and independence, despite the elementary connectedness and unity between the part and the whole. This paradox of simultaneous oneness and difference permeates all levels of reality within the universe. It also stands at the bases of all the mind boggling and brain wrecking philosophical, mathematical and scientific paradoxes.

The paradoxical and seemingly incomprehensible nature of the universe is one of the key reasons why scientists and philosophers have become sceptical and frustrated. Each time a solution presents itself in the form of a comprehensive theory of the universe, there will be an observation or mathematical equation that contradicts the newly found solution. Another reason for scepticism is the incredible complexity and the inconceivable nature of the universe. The fact that we are dealing with something that is essentially infinite, baffles even the most brilliant, yet finite mind. This has led to the general perception that the more we know, the less we understand, and that definitive answers are fundamentally beyond our reach. Scepticism and relativism are not new. Already in the 18th century the Scottish philosopher David Hume presented his philosophy of fundamental doubt, with a capital D. Hume doubted everything, including the doubt itself. Nothing was certain, even the existence of the outside world could be doubted. While philosophers and scientists were initially euphoric about the discovery of the scientific method, and the power of empirical observation, Hume rudely spoiled this party. Sense experience of external objects, Hume argued, is always indirect, and takes place via media such as light and sound. These media are then translated by the senses into a complex, coded process of electro-chemical impulses, which reach the brain through the nervous system. Within the brain these impulses are decoded once again in order to produce a mental image, of which we assume that it corresponds with the external object. That correspondence is an assumption, we can never know for sure. Whether we use our ears, nose, eyes or touch, the process is always the same. Our consciousness never touches the external objects directly. This uncertainty has never been removed completely, but at the time of Hume this argument played a less significant role. The achievements of science and the technological marvels it produced, confirmed the success of the scientific method. In addition science, headed by Newton, produced a worldview that was both intelligible and comprehensible. A universe made up of hard, indivisible atoms that move in accordance with the mechanical laws of motion, against the backdrop of eternal time and infinite space. That is a type of universe that we can visualize and understand. It is a picture that conforms to our daily experiences and common sense. As a result the doubts introduced by Hume became less relevant.

The physics of the 20th and 21st century is confronted with an entirely different problem. There the question is not so much whether the outside world does or does not exists, but whether we can even imagine and visualize the outside world. Relativity and quantum physics reveal a universe that no longer connects to our ability to develop an intelligible image. How do we visualize the time-space continuum, and the fact that time slows down as an object increases its speed. How do we picture a sub-atomic universe, where in a paradoxical manner particles sometimes appear as particles and sometimes appear as an energy field. Even more astounding is how such minuscule particles can generate a destructive force and energy that defies all imagination. The Big Bang is another mystery. An explosion or massive inflation is something that we can still comprehend and imagine. The moment we try to imagine the cause of the Big Bang, however, we enter the realm of the inconceivable and the incomprehensible. According to the prevailing Big Bang cosmology, the Big Bang came out of the Singularity, an infinitesimal small point, beyond time and space, within which an infinite amount of mass and energy is compressed at an infinitely high temperature. Just try to imagine that.

The philosophers are waiting for the scientists to provide answers, while the scientists themselves are plagued by insolvable logical and empirical paradoxes. Their observations no longer provide clarification, as they did in the earlier phase of the scientific revolution, but instead lead to more confusion. Therefore the answers are not forthcoming, and the philosophers can wait forever.

One thing however is entirely clear, a fact that is accepted by scientists, philosophers and sceptics. The universe is incredibly, but incredibly complex. This complexity has an organizational level that is truly astounding. Despite all the paradoxical observations, scientists can agree on one thing without a shadow of doubt. The universe is unbelievably complex, on all levels of existence, from sub atomic structures to clusters of galaxies. This organized complexity forces upon us a question and begs for an answer. Where did this organized complexity come from and how did it arise. There are only two possible answers: it came about by chance or it did not come about by chance. The first answer does not require any further explanation, in the sense that it does not refer to anything else. Chance and chaos stand in and by themselves and are the logical result of a universe consisting of innumerable independent particles and energies that do their own thing, acting without any coordination. The problem is however that detailed mathematical calculations have demonstrated that the complexity of the universe is so incredibly large, that such complexity could never have arisen by chance. Recent insights in the field of mathematics, physics and cosmology in this respect, are truly astounding. These discoveries have revealed that the universe must have been perfectly balanced and fine-tuned from the very first fraction of a fraction of a millisecond after the Big Bang. This is an astonishing and baffling conclusion. The famous mathematician and physicist Roger Penrose has calculated the improbability that this fine-tuning could have occurred by chance at the moment of the creation of a universe. The number that expresses this improbability $10^{10/123}$ is so incredibly large that if you were to write it out in a normal 10 point number type, it would fill up most of the known universe. Here we are not dealing anymore with improbabilities, but with total impossibilities. Atheists defend this impossibility by resorting to exotic, fundamentally unverifiable hypothesis, such as the multiverse theory. If chance is an inadequate explanation, then the answer must be: no chance. The opposite of chance however is not "no chance", but it is intelligence, coordination and consciousness. That is the subject of this book, Intelligence or Chaos, Intelligence or Chance as the only two possible explanations for the existence of organized complexity in the universe. With the conclusion that chance has no chance, and that chance and chaos as explanations for the complexity of the universe, are totally inadequate.

Despite all the "fuzziness" of modern scientific theories, this is an area where science is able to finally formulate a clear and decisive answer. Whether the outside world does or does not exist, whether we can visualize the outside world or not, one thing is clear. Reality is inconceivably complex, and there must be an explanation for that complexity. That explanation forms the demarcation line between theism and atheism, between intelligence and chaos, between an intelligent creative force and a dumb singularity, between consciously orchestrated creation and an unintelligent, dumb explosion. This is finally a question that both philosophers and scientists can provide an answer to, and where no one can hide behind vague scepticism and nihilism.

Besides normal human curiosity there is another reason why the answer to this question is of extreme importance. That importance has manifested itself in our recent world history, and relates to the fact that people always act on the bases of a certain vision and worldview. With the rise of Nazism and Marxism, two repressive ideologies, the world has been presented with the consequences. Driven by the madness of its leaders, the fanaticism of their followers, these ideologies caused World War Two, the biggest and most bloody conflict in human history. Over fifty million soldiers and civilians perished in this insane conflict. World War II is one of the few wars in human history in which the difference between good and evil was so clearly pronounced. It was a war with a high ideological content, beside the normal hunger for expansion and craving for power. The ideological content exposed the influence that an ideology can have on human affairs and behaviour. It also exposed the vulnerability of human societies towards misleading ideologies. The Nazi ideology has demonstrated, to an extreme degree, how Good can be presented as Evil, and Evil as Good. It has demonstrated how compassion can be degraded to a symptom of human weakness, and how cruelty can be elevated to a desirable ideal. It has demonstrated how normal human beings, corrupted by terror and indoctrination, can be enticed to commit the most atrocious acts. In the brilliant but deeply tragic movie "Schindlers List" of Steven Spielberg, this human drama is portrayed in a truly shocking and deeply moving manner. Good can become Evil, and Evil can become Good, to the point that people actually experience it that way.

But also the heinous crimes of Joseph Stalin were justified by ideological indoctrination, supported by unprecedented terror and arbitrariness. His methods of repression and cruelty showed a pattern that was similar to that of Hitler, but was in some respects even more pervasive. The Marxism of Stalin was, on paper an idealistic and benevolent ideology, the so-called workers' paradise. In practice it was a ruthless reign of terror ran by a violent and paranoid dictator. In Stalin's interpretation of Marxist ideology Good and Evil swapped places as well. It is no coincidence that both ideologies directed their hatred and destructive powers towards each other. A deeper analysis reveals that both ideologies were driven by one underlying common worldview. This common worldview is atheistic materialism, with at its cornerstone the theory of biological evolution and natural selection. This worldview implies the relativeness of all life, including human life. It ignores the aspiration for individual happiness, and makes it subordinate to a so called higher goal. This higher goal is evolution, and the perfection of the species through the struggle for survival and natural selection. The cruelty of nature and the animal kingdom exemplify this process, and form the guidelines for the norms, values and morality of human civilization. The weak are being conquered and subjugated by the strong, who are free to act as they wish. The weak are enslaved, exploited, tormented and then devoured. This principle is glorified, and the pain and suffering of the weak are similarly glorified. Compassion is seen as weakness and cruelty is seen as a virtue.

Theistic ideologies too have been guilty of violence and cruelty. The moment fundamentalist interpretations of a theistic ideology gain the upper hand, such an ideology becomes dangerous. There are plenty of examples of this. In this situation too, a lack of compassion combined with an ideology where the 'goal justifies all means', leads to a reversal of good and evil. Atheistic ideologies find justification of cruelty in relativizing human suffering, the supremacy of the struggle for survival and the principle that might makes right. Cruelties committed in the name of corrupted theistic ideologies have a different cause. There the cruelties are justified by glorifying human suffering as atonement for committed sins. In the middle ages the Spanish inquisition condemned so called witches to a cruel death by being burned at the stake. The perverse rationale was that the suffering in this life by being burned at the stake, would save the sinner from burning in hell for all of eternity.

Worldviews form the foundation upon which human society and civilisation is build, and their influence is not to be underestimated. Worldviews are the operating system that runs the software of our values, norms and actions. Every human being must take decisions, every single day. Every act and decision is based on a vision and a worldview, consciously or sub-consciously. The worldview is the filter, the coloured spectacles through which we view reality, often times forgetting that we are wearing spectacles. When a worldview influences large numbers of individuals, such a worldview will define the nature and quality of a human society. In the end there are only two operating systems in the world that contain an opposing set of instructions: atheism and theism. The atheistic universe is ruled by chaos, arbitrariness and violence, flanked by its merciless servants of evolution and natural selection. Chaos, randomness and violence are its core principles. From the very beginning of the universe, from the Big Bang till its final destruction, the universe is undergoing a continuous process of violent change. From exploding super nova's, all devouring black holes, crashing galaxies, or dying stars, it is a continuous orgy of violence of unimaginable proportions. Then, wherever life appears, such as on this planet earth, it reflects on a micro level the violent nature of the macro universe. Life is governed by the vicious struggle for existence and survival. The suffering of one creature leads to the survival and the limited happiness of another creature. Humanists are trying to cover this ruthless struggle with a veneer of human dignity, which dignity is based on nothing. In an atheistic universe there is no dignity and there is never any individual justice. Good and evil are just relative concepts and are subordinate to the laws of evolution and natural selection. And the final goal can never be good; the universe and all the creatures therein shall end in violent destruction, whether that happens through the Big Rip or the Big Crunch.

The theistic universe looks quite different. The foundation of the theistic universe is not violence, but love, compassion and harmony. The theistic universe is ultimately ruled by an infinitely intelligent Being that is in harmony with itself and with its creation. In a theistic universe happiness and harmony are the end goal for all beings. That may sound utopian, but it is what everyone, deep inside, desires. The crucial question is, where is that universe to be found, and why is our universe so totally different. The universe we know is far from the paradise like creation that one might expect from a benevolent, compassionate, almighty, and infinitely intelligent creator. There is an explanation for this which receives extensive consideration in my book. In essence, and that is the view of the Vedanta philosophy, the suffering we experience is caused by the misuse of our minuscule independence, our free will. In fact, this entire material universe is itself the result of the misuse of our free will.

The Vedanta philosophy explains that this material universe is just a part of the totality of creation, and that this particular universe has been created with the specific purpose to satisfy our desire for independence. This is the tension in the relationship between the part and the whole, the delicate balance between harmony and disharmony, and where our free will tips the direction of the scale.

When theistic ideologies get corrupted and succumb to violence and cruelty, it is without exception caused by a fundamentalist interpretation of its ideology. It can be argued that the cruelty of atheistic ideologies is similarly the result of fundamentalist interpretation of evolution and materialism. To a certain degree this is correct. Humanists will argue that the cruelty of nature and the process of natural selection are not necessarily examples to be followed by human beings. Indeed, there is no reason why human beings should actively participate in applying the process of natural selection, so called to 'lend nature a helping hand'. A broad minded interpretation of natural selection can lead to a situation whereby compassion, altruism and cooperation, will result in a better chance for a group to survive. That too is correct, but that will inevitably be at the expense of another group that is less efficient and less organized. The materialistic universe, driven by evolution and natural selection, is violent in its core. Altruism and compassion are exceptions and are incidental. Therefor it is only a matter of time before a new ideological variation presents itself, in which materialistic atheism and the laws of natural selection are interpreted in a fundamentalist manner. Nazism version 2.0 or Marxism version 3.1.

In the end it is crucial that a higher ideal is applied in the right manner in human society. In such a way that the society flourishes and people in society become happier. There is an old saying, which states that "the path to hell is paved with good intentions." This saying can also be applied to beautiful ideals. The guiding principle for the application of any ideal is the law of karma, the law of universal reciprocity: "Do unto others as you would have them do unto you". The law of Karma is a universal truth, often called the "Golden Rule", and forms the core of every religious tradition in the world. It is found in the Veda's, the Bible and all major religious scriptures. Jesus stated 'Love your neighbour as yourself" and the rules of Dharma in Vedic philosophy, enunciate the same principle. The essence of the law of karma is that one should not cause unnecessary suffering to other beings. If you violate this law, you will have to suffer the same pain you have inflicted on others. Moreover one has a duty, within his means, to help others wherever he can. In this way he secures his own happiness and that of others.

Regardless of whatever ideology one adheres to, this should always be the guiding principle. The problem is that atheistic worldviews do not concern themselves with the laws of karma. In the atheistic universe there is no justice and reciprocity. There exists merely the capricious unpredictability of the laws of evolution and natural selection, directed by chance, randomness and violence. In fact, the fundamentalist interpretation of evolution and natural selection, is nothing but the consistent application of the basic principles of atheism and materialism. It is for that reason that the fundamentalist interpretation of atheistic ideologies is especially dangerous. Such interpretation is intellectually consistent and correct, while the 'extenuating' humanist interpretation is inconsistent, incorrect and therefore hypocritical. For that reason Nazi's and Stalinists felt no remorse when driving innocent people into the gas chambers or the gulags. In their eyes, and the eyes of their leaders, they were acting correctly. They viewed their task as a difficult, but necessary sacrifice, aimed at "lending nature the helping hand" in the process of natural selection. This was particularly the view of the Nazi's.

A good example of this is Stalin's interpretation of Marxist ideology. Historians have concluded that Stalin's extreme cruelty was particularly caused by the fact that Marxism was bereft of any tradition or any 'extenuating' social frame of reference. Even the autocratic tsars that ruled Russia before the communist revolution of 1917, had their aristocratic traditions, ethical norms and hence limitations. Marxism rejected all such traditions as 'bourgeois' and was therefore completely free in choosing a new set of values. Stalin seized this absolute freedom to define his own set of values. Values that suited him, but that, in the name of Marxism, brutalized the Soviet population.

The quality of human civilization is defined by its ability to distinguish between good and evil. In the end this dividing line is determined by the two world views; atheism and theism. The good universe versus the bad, violent universe. To be clear, this does not imply that atheists are bad and theists are good. This analysis is of a strategic nature. It provides the fundamental direction of human civilization, whereby theism and atheism will eventually develop in opposite directions. Also their norms and values, the criteria of good and evil, will develop in opposite directions. Atheism relativizes life and the individual pursuit of happiness, because there are no absolute values. The suffering of one creature can be the limited happiness of another, and 'one man's death is another man's bread'. For the individual however, good and evil are not at all relative concepts. The suffering that we experience as an individual is real, regardless of its cause. Nobody wants to suffer, every being searches for happiness. 'The pursuit of happiness' as it has been so eloquently described in the American Declaration of Independence, is the driving force behind the actions of every being. This is also the essence of the law of Karma, the right of happiness for every individual. In the end a worldview and ideology must serve this purpose and nothing else.

Thereby it is of course essential that the philosophical conclusions of a worldview support these ethical considerations. I am of the opinion that the atheistic worldview is both scientifically and philosophically incorrect. The evidence for a universe guided by intelligence is quickly accumulating. At the same time chaos and chance, as ultimate causes, are becoming increasingly improbable. From a mathematical standpoint they have evolved from improbable to impossible. That is an objective observation, a hard scientific conclusion, that is not determined by personal preferences or ethical, moral considerations. Beside its scientific and philosophical imperfection, nihilism and materialism will lead to dissatisfaction and desperation on an individual level. No one can flourish in an environment of pointless futility. It is the nature of human beings to want to understand and to search for meaning. Consciousness is in its core a meaning searching and meaning creating phenomena. It is significant that in the developed world, where materialism and atheism are widespread, many persons will still claim to believe in 'something'. This 'Something-ism' has practically assumed the proportions of a new religion, and demonstrates the desire for understanding and meaning. In the hands of fundamentalist fanatics, the materialist and atheistic worldview can be a fertile ground for destructive ideologies. The ideologies that have created World War II are a painful example of this. At this time of global communication, of unbelievable efforts in the field of science and technology, but also of weapons of mass destruction, knowledge and understanding are more important than ever before. This book is a humble effort to contribute to this.

September 2021

Chapter 1 – The mystery of existence.

This chapter deals with the existential questions of life, why do we exist, why does anything exist and is there a meaning to existence other than existence itself. The chapter starts by describing the surrealistic position in which we find ourselves; our daily, earthly reality against the backdrop of an inconceivably complex, infinite and mysterious universe. Moreover, our earthly reality is far from perfect, where we are all confronted with different degrees of suffering, man-made such as foolish wars as well as suffering from natural causes. Inevitably and inescapably we are all confronted with death. Against this backdrop a human being naturally wonders about his existence and its meaning. The issue of meaning is directly connected to the central theme of this book and the most important question that humanity can ask itself, whether or not God exists. The atheism versus theism debate is immediately placed to the foreground, whereby atheism is defined by four key assumptions and propositions:

1. The universe consists of independently and chaotically operating particles of matter separated by empty space. Any appearance of structured complexity and design is

caused by chance, emerging from fundamental chaos, and not by a coordinating, all-encompassing intelligence, such as God;

2. The universe only consists of matter and void, and there is no other energy or reality outside or beyond matter

3. If the universe had a beginning or cause, this cause has to be of ultimate simplicity. Since God is a complex being he could not have been the ultimate cause of the universe, since he would have to have been created by something else,

4. Life as we know it on this earth, is temporary and full of useless suffering. If God were all-powerful and good, he is either a sadist or incompetent. More likely he does not exist.

While atheists acknowledge the existence of a number of natural laws and mechanistic forces, such as Newton's laws of motion, these laws only produce predictable results when their mechanisms and interactions are applied to a small number of particles. On a large scale these mechanistic forces result in random interactions, devoid of central coordination, which is the reason the universe is labelled by materialists as the "Random Universe".

Chapter 2 – Intelligence or Chaos – the teleological argument.

This chapter deals with the first atheist proposition, and is countered by the teleological argument, the argument from design. While the teleological argument has been around for centuries, it is today strongly supported by scientific evidence, based on the fairly recent discovery that the universe is fine-tuned. This fine tuning comprises the four basic forces, gravity, electromagnetism, and the strong and weak nuclear forces that regulate everything that happens in the universe. These four forces are constants, i.e. they are causeless, but mysteriously happen to have the exact right values in relation to one another to enable the existence of the universe; from the existence of sub-sub atomic particles to the existence of mega clusters of galaxies. Moreover these forces and their values were created instantly at the first few seconds of the Big Bang. The probability that this could have happened by chance at the time of the Big Bang is statistically beyond impossible, as expressed in the number $10^{10^{/123}}$. The conclusion is that the universe must have been created and is being maintained by a transcendent and intelligent being, God. The teleological argument is also extended to the emergence of living organisms, and the statistical impossibility that the inconceivable complexity of just a single living cell, could have arisen out of random combinations of atoms and molecules. The chapter ends by describing the only possible rescue for atheism, which is the multiverse. This lifeline is illusory though for it refers to the existence of fundamentally unverifiable realities, such as parallel universes that exist in entirely different realities and time-space continuums, totally disconnected from our reality. As such they can never be observed, not even in principle, and are therefore meaningless from a scientific perspective.

Chapter 3 - Materialism – Matter is the only reality.

This chapter deals with the second atheist proposition, which states that matter is the only reality in the universe, or materialism in short. First an explanation is given on materialism, and how it has evolved into the modern scientific narrative of reality. It then traces this back to the beginning of the scientific revolution, and the apparently logical progression towards materialism dictated by a number of successive scientific discoveries and the resultant theories they produced. In particular three scientific theories were revolutionary and have altered the course of science and history, Newton's laws of motion, Mendeleev's periodic table of chemical elements, and Darwin's theory of biological evolution. Each one, particularly evolution has contributed to the ascendance of mechanistic materialism and in its wake, atheism.

Chapter 4 – Darwin's theory of evolution.

This chapter deals exclusively with Darwin and biological evolution, since of the three scientific discoveries, evolution has had by far the most impact on the way we look at the universe and our position therein as human beings. Many atheists see Darwin's theory as the completion of the atheistic worldview and in the words of Dawkins "Darwin made it possible to become an intellectually fulfilled atheist." The core tenets of the theory are discussed, as well as criticism that has been levelled against the theory by various scientists. One major criticism is based on the teleological argument, mentioned in Chapter 2, other criticisms include the incomplete fossil record, the flawed mechanisms of genetics, the problems with mutations, the inability to explain the origin of life and archaeological evidence that significantly contradicts the evolutionary time line. The impact of Darwinism on society and politics, also called social Darwinism, are being analysed, and its impact has been enormous during the 20th century. Both Nazism and Stalin's brand of communism were heavily influenced by social Darwinism

Chapter 5 – The defects of materialism.

This chapter discusses all the flaws and limitations of mechanistic materialism, concluding that materialism is an incomplete and fundamentally flawed worldview. The advent of the New Physics, including relativity, quantum theory, Big Bang Cosmology, the Anthropic Principle and the Cosmic Fine-Tuning all point to a mysterious and energetic universe, not driven by mechanistic principles and not limited to the mere existence of matter. Before Einstein energy was considered to be an extension and by-product of matter, after Einstein and his famous equation $E=MC^2$ matter turned out be a by-product and transformation of energy. Energy in many ways represents an entirely different reality, manifesting itself as energy fields and energy forces that operate on the bases of quite different principles, such as information, creation, synchronicity, and intelligence. Quantum field theory specifically encompasses such a reality, thereby ascertaining that reality is fundamentally paradoxical as exemplified by the wave-particle duality. This theory recognises on the bases of empirical observation that material particles are simultaneously particles and energy fields, two mutually exclusive states of being. This chapter further deals with the problems and limitation of sense perception, relativity and quantum field theory, the existence of dark matter and dark energy, and the implications of the Big Bang and the singularity from which the Big Bang has arisen. All of these undermine the mechanistic, materialist worldview and point to a universe that's driven by information and consciousness. The conclusion of chapter 5 is that the second atheist proposition is incorrect and that it has been thoroughly refuted.

Chapter 6 – The origin of existence has to be 'simple'.

This chapter deals with the third atheist proposition that if the universe has a beginning and a cause, such a cause must be of absolute and ultimate simplicity. Dawkins argues this position as an intuitively felt axiomatic truth. The issue has become extremely relevant since science has embraced the Big Bang as the likely explanation for the origin of the universe, and accepted the notion that the universe had a beginning. Further explanation is provided on the Big Bang and its competitor the Steady State theory, which theory assumes the universe is eternal, fundamentally unchanging and stable. In the course of the 20th century the Steady State theory was abandoned by scientists, in favour of the Big Bang. The importance of the Steady State theory is further discussed in terms of its attractiveness to materialism and atheism, and its fundamental persuasiveness in this respect. Even Einstein was an adherent of the Steady State theory until Hubble and Lemaitre convinced him otherwise.

The counterargument to the notion of a cause of ultimate simplicity, is that any existence by its very nature cannot be just simple. This point is further explained in Chapter 9. Even the singularity, as the name suggests supposedly the simplest conceivable cause of a complex universe, fails in this respect. For the singularity is described and defined as an infinitely small point that contains an infinite amount of compressed mass and energy, at an infinitely high temperature. Anything that possesses these attributes does not qualify as ultimately simple. The only thing that is ultimately simple, is nothingness. Since nothingness does not exist, it cannot have any effect on existence. Therefore the cause of the universe may well be ultimately simple, but must simultaneously contain ultimate complexity, otherwise the entire range of properties and complexities that make up the universe, would have to have emerged from nothingness. This issue directly relates to the law of causality, of cause and effect, which states that every effect must be contained within its cause, actually or potentially. The law of causality also relates to the three laws of logic, the law of identity, the law of non-contradiction, and law of the excluded middle, which are explained in some detail. The laws of logic are based on the stringent distinction between being and absolute non-being, and the fact that absolute non-being does not exist. In the end atheism puts nothingness forward as the ultimate explanatory principle, suggesting that the singularity and subsequent big bang emerged out of pure nothingness. The theist position is the exact opposite, the universe and its entire range of properties and complexities has emerged from the infinite fullness of being, which is precisely one of the definitions of God.

Chapter 7 – The imperfect universe.

This chapter deals with the 4th atheist proposition, which is the problem of evil, also called the theodicy. Many philosophers consider this the most important argument presented by atheist against theism and the existence of God. If God is all good and all powerful, how can he create a universe that is, from a human perspective, highly imperfect and full of unnecessary and undesirable suffering? This topic is of a more philosophical rather than a scientific nature, since it deals with the way living beings experience the universe from a psychological perspective. To answer this question reference is made to the Vedanta philosophy, a 5.000 year old school of thought from ancient India. According to Vedanta philosophy every living being is part of God's internal spiritual energy, and therefore possessed of consciousness, personality and individuality. As conscious beings we have a choice to either cooperate with the cosmic order, of which we are a part, or to desire greater independence and greater control. When the desire for independence and power exceeds our feelings of love and service, a living entity is transferred from the spiritual realm to a different reality and a different universe, the universe made of matter. This is the universe in which we presently reside. The material universe is basically an alternative, illusory reality, in which a living entity can exercise his independence to different degrees. Since power has an intoxicating effect, such exercise can lead to the abuse of power, negatively effecting the lives of other living beings. Such abuse invokes the laws of karma and hence individual suffering is created.

In this chapter the laws of karma and reincarnation are further explained according to Vedanta philosophy, whereby the point is made that without understanding these laws, theism becomes indefensible. It would relegate suffering to the arbitrary whims of God, dished out to individuals within just one life, without any connection to cause and effect. The law of karma states that suffering experienced in this life is the result of previous good or bad actions committed by a living being, either in this life or in previous lives. Therefore every individual reaps what he has sowed and is responsible for his own actions and its consequences.

Chapter 8 – Why is there something rather than nothing?

This chapter continues on from chapter 6, and more specifically deals with the issue as to why there is something rather than nothing, and the relationship between being and non-being. An important distinction is made between absolute and relative non-existence. This chapter clarifies how atheism relies on the concept of absolute non-existence to explain the origin of the universe, as well as to explain the principle of evolution, and not just the Darwinian version of evolution. For the evolution principle implies, as the term suggests, that in due course of time entirely new properties emerge out of an existing object. These entirely new properties must necessarily emerge out of nothing, since they were not present in the original object. It is then explained how and why this interpretation of the evolution principle is incorrect, for simply put, something cannot come from nothing.

This chapter further deals with the notion of the eternal co-existence of atoms and void, or being and nothingness, as embodied in the Steady State theory referred to in Chapter 6. According to this theory the universe as we know it is eternal, whereby material particles and the empty space in between these particles, co-exist eternally. The Steady State theory has been both scientifically and philosophically refuted, scientifically by the Big Bang cosmology, Einstein's theory of Relativity, and Quantum Field theory which states that space is not empty but full of quantum energy. It has been philosophically refuted because absolute non-existence by definition cannot exist. The second part of the chapter deals with and refutes the notion of subjective idealism or solipsism, the notion that external reality is created by individual consciousness. Arguments are then put forward to support the theory of objective idealism, which assigns creation and intelligence to the existence of an objective conscious agent, i.e. God. The last part of the chapter addresses the issue of effectiveness within the context of an atheistic universe, questioning why the universe would want to create itself in the first place. In quoting Stephen Hawking: "Why does the universe go through all the bother of existing?" the argument is made that non-existence is more effective than existence.

Chapter 9 - Why does anything exist in this particular manner?

This chapter explains that being in and of itself is by definition complex, which complexities are irreducible realities. Being is analysed and dissected on a most fundamental level, that reveal that there are inalienable properties that inherit in all forms of existence without which existence would not qualify as existence. These are mass, energy, attributes, motion, change, time, space, organized complexity and consciousness. In addition there is the internal reality of the conscious observer that also entails inalienable properties, such as intelligence, emotions, will power, desire and self-consciousness. The chapter starts with the question why reality is the way it is, which question is answered by referring to the inalienable properties that make existence into existence, and without which existence could not exist. These truths are self-evident and refer to the most basic and universal principles of empirical observation. At the end of the chapter all the inherent complexities of being, such as mass, energy, properties, time, space and consciousness are briefly described.

Chapter 10 - The ultimate complexity, consciousness.

This chapter refers to the fact that of all innate complexities, consciousness is the most extra-ordinary, unique, and the most important one. Consciousness is distinguished from Artificial Intelligence (AI) and computer generated information, whereby 'understanding' the 'conscious experience' and the presence of 'emotions' are the most distinguishing features. Consciousness is then analysed into its constituent components such as passive and active, which are then subdivided into intelligent awareness and emotional awareness on the passive side, and desire and willpower on the active side. The mysterious nature of consciousness is explained and how it is fundamentally paradoxical. Consciousness is simultaneously one and many, simultaneously full and empty, simultaneously changing and permanent, and simultaneously able to generate action within inaction. It makes consciousness into a reality that is in a completely unique class of its own. These paradoxical features of consciousness, such as simultaneous oneness and manyness, stand at the bases and form the core of what we call information. All information is based on a relationship between the part and the whole and the essential oneness and difference this relationship implies. For instance in mathematics numbers are only meaningful because they are part of a set of numbers. The number 2 derives its meaning from being preceded by number 1 and succeeded by number 3. Most importantly information is meaningful on the level of consciousness because a diversity of sensations is merged into a single conscious experience, experienced by the I, the self, the conscious observer. Thereby the parts and the whole, and the relationship between them, are simultaneously perceived leading to what we call and experience as 'understanding'.

Chapter 11 - God and Reason.

This chapter describes the problematic relationship between religion and science, and how the western religions have struggled with this phenomena. The subordination of empirical knowledge and logic to revealed scripture, has been especially problematic. In the words of Tertullian, an important and influential 2nd century Christian Church Father: "I believe because it is absurd". And in the 16th century Martin Luther, the founder of the protestant reformation stated: "Reason is the whore of the devil". The embrace of the absurd and of irrationality has produced many conflicts especially when the scientific revolution started taking shape in Europe in the 15th and 16th century. The trial of Galileo and the execution of Bruno Giordano were historical lows in the history of Christianity and have highlighted this area of extreme tension. Vedanta philosophy to the contrary accepts empirical evidence and logic as valid means of acquiring knowledge, in addition to the mystic experience and divine revelation. The chapter ends by emphasizing that the convergence of science, philosophy and religion is necessary in order for mankind to achieve a higher level of comprehensive knowledge and understanding. In this chapter arguments are put forward for the universality of religious and spiritual principles, denouncing sectarianism and fundamentalism.

Chapter 12 – The paradox; the foundation of reality.

This chapter is in some ways the quintessence of the entire book, because it lays bare the paradoxical and inconceivable nature of reality. The core paradox entails the simultaneous oneness and manyness of God and his energies, described by the great mystic and philosopher Caitanya Mahaprabhu as "Acintya bheda abheda tattva". This Sanskrit term literally means that reality, truth, consists of simultaneous and inconceivable oneness and difference. In the book this principle is referred to as the Acintya principle. While Acintya refers to inconceivability on a logical plane, emphasis is made that Acintya is not irrational but supra-rational, because it can be empirically verified. To that extent many practical examples are given of the Acintya principle in ordinary and scientific experiences. A brief summary is provided of the history of paradoxes, whether logical, mathematical or empirically, and the distinction between the supra-rational and the irrational is further explained. The wave particle duality in quantum physics is supra rational, whereas a square circle is irrational. Further explanation is given about the inconceivable nature of God, as the embodiment of the Acintya principle. Reference is made to the paradoxical properties of consciousness as described in chapter 10, and these are extended to the nature of ultimate reality itself, pure energy. These paradoxes that are based on everyday empirical observation, are then discussed one by one.

Chapter 13 – The nature of God.

In this chapter further detail is provided on the nature of God according to Vedanta philosophy, specifically whether God is personal or impersonal. In the Indian tradition this subject has been a source of fierce debates as well as disagreements. The distinction is directly related to the difference between monism and monotheism, as briefly described in Chapter 1. Vedanta philosophy in its core is monotheistic and personalist, whereby God is both personal, localized and transcendental, and is yet simultaneously impersonal, non-localized and all-pervasive. However, there is a hierarchical relationship between the personal and the impersonal, whereby the personal is of the highest order. The relation of the personal to the impersonal also refers back to the Acintya principle described in the previous chapter. The personalist school is called Vaishnava Vedanta, as opposed to the Advaita Vedanta school of Sankara. The monotheistic philosophy has a direct impact on our position as living beings. According to monism, as expounded for instance by Sankara, living beings are not just part of God, they are one with and identical to God. Once a living being is liberated from Maya, or illusion, he merges with God, loses his individuality, and becomes one with God. Vaishnava Vedanta rejects the notion of merging with God and a loss of individuality. Rather liberation implies the rediscovery of one's eternal and spiritual individuality in full consciousness of God, and in an eternal loving relationship with God.

The chapter then goes on to explain the difference between conditioned, material consciousness, and expanded spiritual consciousness. Expanded consciousness enables a finite living being to reconnect with the internal energies of God, and tapping into the infinite reservoir of information, knowledge and blissful emotions. In this state the enlightened living being is able to experience a range and flavours of blissful emotions he did not even know they existed. In such a state of permanent fulfilment a living being experiences an unimaginable state of satisfaction and happiness, something every living being is always searching and craving for in this world.

Chapter 14 – Vaikuntha, our final destination.

This chapter presents the conclusions and summary of the entire book, which are summed up in four statements. These conclusions are based on the refutation of atheism as presented in the first seven chapters. To that is added the essence of Vedanta philosophy and its conclusions. The four conclusions also correspond in some ways to the four noble truths of Buddhism, and are stated as follows:

1. God exists and the material universe in which we live, is part of his creation. We, as living beings, are an eternal integral part of God, qualitatively one yet quantitatively different. As parts of God we share the same properties of

Sat, Chit Ananda, or eternal existence full of knowledge and bliss.

2. Life in this material world is not an evolutionary accident, but has purpose and meaning, even though it is characterized by suffering, such as old age, disease and death. Suffering is caused by karma that we ourselves have generated in the past.

3. By the exercise of our free will we are able to choose a path that can lead us to liberation from this material world. Throughout the centuries religions in their own specific way, and in accordance with time, place and circumstances, have provided guidelines as to how we can progress spiritually, and how we can liberate ourselves from the bondage of the material world. The Vedic system employs the various yoga systems to achieve this goal.

4. There is another, spiritual universe where the limitations and imperfections of material existence are not present. The Veda's call this universe Vaikuntha, which literally means freedom from anxiety. The final destination of every living being is to return to Vaikuntha.

The chapter then goes on to describe Vaikuntha, what it's made of, what it looks like and what the conditions are, specifically in contrast with the world we know, the universe made of matter. It explains how the material universe is in effect an upside down, perverted reflection of the spiritual world, which the Veda's compare to a banyan tree. Nonetheless, as a reflection it also possesses similarities from which we can deduce the nature and properties of the spiritual world. The structure of the spiritual world in this respect resembles the material universe, with planets and galaxies, located in the infinite and luminous firmament, the Spiritual Sky and inhabited by innumerable living beings. It further elaborates on the state of expanded consciousness of its inhabitants, and their interactions. A perverted reflection also implies major differences compared to the original, such as the absence of birth and death, the absence of any suffering and a perpetual state of bliss. Moreover the spiritual world is luminous and possessed of an inconceivable beauty. Most importantly the presence of God is everywhere, where living beings and God are engaged in eternal pastimes and entertainment. Since the spiritual world is perfect, there is no work to be done and no mission to be accomplished other than its inhabitants engaging in playful, loving exchanges, surrounded by astounding and indescribable beauty, harmony, intimate cosiness as well as majestic splendour.

> *"The most incomprehensible thing about the universe is that it is comprehensible at all"*
>
> **Albert Einstein** [1]

Unimaginably large numbers!

When I look out of the window of my study to the world outside, I see the world as we know it. I see trees, gardens and buildings bathing in bright sunlight, except for the shadow of the occasional cloud passing overhead. Around me, life is taking its course. Nothing remarkable, as you might say; everything is just as we know it. However, behind this everyday reality exists a universe of an almost unimaginable size and complexity. We can forget this universe so easily in our daily routine of work, grocery shopping and enjoying our free time, but it is nevertheless always present, just behind that blue or cloudy sky, and it is full of truly astonishing phenomena. Take the mysterious and intriguing phenomenon of light for example, which makes all life on this planet possible. It is only because of light that we can actually see anything of the world around us, yet rarely do we consider that this light has just made a vast cosmic journey simply to get here. Emanating from what we call the sun, a relatively small star known as a 'yellow dwarf' in astronomer's jargon, the light that reaches our planet earth has travelled 150 million kilometres at a speed of about 300,000 kilometres per second taking roughly just 8 minutes to complete the journey.

The sun may be small compared to other stars, but the force that she produces is still unimaginably powerful. Every second our star produces an amount of energy that equals the explosion of 1 trillion 1 megaton hydrogen bombs. In this same second, the sun produces enough energy to keep the entire world economy going for 500,000 years based on our current energy usage. Due to the enormous amounts of energy being produced and the speed at which it travels, we can feel the influence of the sun

almost immediately despite her distance from earth. On a hot summer's day, her heat can be unbearable and we are grateful just to find a spot in the shade.

However, the sun is only a glowing pin-head compared to the total size of the universe. To be more impressed by the cosmos, we have to wait until the sun disappears behind the horizon and darkness sets in. After sunset, the scale of the universe becomes more apparent as numerous stars, star systems and other celestial bodies appear in the night sky. For those of us not living in towns and cities and not hindered by light pollution the night sky would be filled with thousands of stars. Nonetheless, no matter how impressive a view, we would only be witnessing a tiny part of the entire universe, a fraction of a fraction of an immeasurably large space.

Those who want to have a deeper understanding of the universe will have to enter the domain of extremely large numbers. The distances within our own solar system are already enormous. Earth is part of a collection of nine planets, including the lonely outer dwarf planet Pluto. The distance from the sun

Our solar system with its 8 planets Mercury, Venus, Earth, Mars, Jupiter, Saturn, Uranus and Neptune.

to Pluto is, on average, 5 billion kilometres. If we were to travel by spaceship at the impressive speed of 60,000 kilometres per hour, we would have to travel non-stop for 10 years to cover this distance. But if we zoom out further, our solar system is gradually reduced to a mere dot in the vast expanse of space. Our collection of planets is a miniscule part of a much larger entity; a galaxy called the Milky Way. The distance from one side of this system to the other side is 100,000 light years. One light year is the distance that light travels in one year at the speed of 300.000 km per second, or 9.4 trillion (9,400,000,000,000) kilometres. If we continued to travel in the spaceship that took us to Pluto at the same speed, it would take us 1.8 billion years to traverse the Milky Way from one side to the other.

Nevertheless, we would still be 'safely' within our own star system. If we ventured to travel to our next nearest major

The Andromeda star system is located at a distance of 2.4 million lightyears from our solar system.

star system, the Andromeda galaxy, we would have to cover a distance of 2.4 million light years. If we continued to travel at this same speed of 60.000 km per hour, it would take us no less than 43.2 billion years to get there! These distances are totally beyond human comprehension. We can hardly pronounce such numbers, let alone imagine them. Who does not, from time to time, look up to the stars in the sky and wonder with amazement and apprehension where it all ends?

Compared to the stupendous size of the universe this planet earth and its inhabitants shrink to insignificance. This planet is inhabited by a vast array of living creatures, from bacteria, plants, animals to humans endowed with intelligence. In this theatre of living creatures humans occupy a special place, because they are able to control, manipulate and organize their environment. Human beings build civilizations, creating order in a hostile environment, aimed at survival and protecting life and livelihood. As Maslow has eloquently explained, once these basic needs are fulfilled, civilisations broaden their goals to improve the quality of life, by the cultivation of knowledge, science and art. Overall, looking at our historical records, we have not been very good at this, mostly hovering at the lower end of Maslow's pyramid. Civilizations have been primarily focused on survival, protection and defence and a lot of knowledge has been cultivated with military objectives in mind. According to a New York Times article published on July 6 2003, over the past 3,400 years humans have been entirely at peace for just 268 years, or just 8% of recorded history. That means there were wars going on for 3.132 years somewhere on the planet. These wars have claimed between 150 million to 1 billion casualties. That's not a very good statistic, and it says a lot about the human condition and the quality of our civilization. The relative peace of the past 70 years is mainly due to the existence of nuclear weapons, which make it impossible for us to have large scale wars. While most wars, in hindsight and almost without exception, seem to be useless, a nuclear war is useless in advance. The so-called MAD doctrine of Mutually Assured Destruction is an insurmountable obstacle to any potential aggressor based on even the most primitive calculations. Looking at our own history, therefore, we mostly seem to be specialized in warfare and fighting one another. Just switch on the news any day and we will be faced with the crude facts of our earthly existence.

This is not minimize the many great efforts that have been made by humans to advance knowledge and eradicate ignorance, to cure diseases, advance social justice and prosperity, as well as shield humanity from hostile weather conditions and natural disasters. Yet, the overall picture is not rosy, and more often than not human beings are seen struggling to fight their complicated battles against real and alleged enemies and threats.

These battles are fought with intense emotions and are of vital importance to each individual. But placed into perspective, these great and small human activities take place against the backdrop of nature and the infinite universe. Only one hundred kilometres of atmosphere separate us from the unreal reality of this immeasurable, unimaginable universe. These one hundred kilometres above our earth are the boundaries of the tiny bubble in which earthly existence takes place. This tiny bubble, earth and its atmosphere, floats in an immeasurable ocean of cosmic energies of outright extra-terrestrial proportions.

The difference between the immeasurable universe and daily human life and worries is surreal. It is a remarkable contrast; the cold, uninterested magnitude of the universe set against the intense emotions and awareness of our miniscule existence, occurring simultaneously and of course, both equally real. But, what is 'real'? Why does reality exist? Just like everyone looks at the stars now and then and wonders about the vastness of the universe, everyone will sooner or later also wonder why we exist and why everything around us exists. What is remarkable is that reality appears to be inflexible and it does not seem to be interested in us or our well-being. Both the world and the universe just exist, distant and indifferent; at least so it seems. In the words of Richard Dawkins:

> *"The universe we observe has precisely the properties we should expect if there is, at bottom, no design, no purpose, no evil no good, nothing but blind, pitiless indifference."*[2]

This nihilistic explanation causes problems and raises questions. We wonder, why is there something instead of just nothing? Why does reality exist the way it does? Why is reality at every level so immensely complicated? And why is reality permeated with undesirable things such as old age, disease and death and other types of suffering. Is there an explanation other than the one provided by Dawkins above, or is that it.

These questions form the foundation of this book. They are

the starting-point towards the question that defines the mystery of existence and that is the most important question that human beings can ask themselves: Does God exist or not? Does existence – small or gigantic – spring from an unconscious and unintelligent chaos, or is it created by consciousness and intelligence and does it have a purpose and a design? The answers to these questions will provide an insight into our role and position in the universe. Do our lives have meaning or is our existence totally lacking any purpose? Do human beings exist with an intention, or do we just float around in the cosmos without ever achieving anything? Or, as the famous atheist philosopher Bertrand Russell put it so strikingly:

Thanks to: NASA, ESA and the Hubble Heritage Team STScI/AURA. The Andromeda system is situated at an impressive distance of 2.4 million light-years away from us. This distance is nothing compared to the distance to NGC 1300, a spiral-shaped star system that is situated at a distance of 61 million light-years away from us in the Eridanus constellation. The star system has a cross-section of about 110,000 light-years; just slightly bigger than our own Milky Way.

> *"Man is un unfortunate accident, a sideshow in the material universe – an odd accident in a forgotten corner."* [3]

Ultimately, we are of course all interested, out of normal self-interest, in our own position and perspective in life. At the deepest level, this perspective is completely determined by the answer to the question whether God exists or not.

Philosophical analyses shows that the answer to this question will determine the way we view reality and the universe. All philosophies can, in the end, be divided into two fundamental categories. The first category is atheistic in its core and states that the origin and the functioning of reality is based on chaos and coincidence. The second category is theistic in its core and regards the universe as an organic reality that was created and is managed by intelligence. Other philosophical systems that are essentially agnostic – and therefore do not explicitly state whether God exists or not – are often considered to be atheistic. In most cases, they will state that the intelligent coordination of the universe is an improbability. Therefore, they implicitly – and based on elimination – have a preference for chaos and chance as the most probable explanation for the origin of the universe.

Of course, within each of these categories there is a great diversity of philosophical systems with many differences in nuance. Nonetheless, the dividing line is striking and this has an all-determining effect on all aspects of a philosophy, such as the theory of knowledge (epistemology), the theory on the nature of being (ontology), theories concerning moral values and meaning (ethics) and, in the end, the description or perception of our physical and scientific reality (physics and metaphysics). Indeed, social and political ideologies are also largely defined by this split. Denying or confirming the existence of God therefore leads to opposite philosophies and completely opposite answers as far as the origin and meaning of existence is concerned. Do our lives have a deeper meaning, or are our lives meaningless; a random evolutionary accident? Is man just a product of matter, or is there another type of energy that defines our consciousness

The images of this rich set of star systems are made by La Silla Observatory of the ESO in Chile. The thousands of star systems that are situated in this small area of the firmament provide us with a look into the distant past of the universe and makes us realise again how enormously large the cosmos is. Just underneath the bright stars in the centre of this image there is a group of star systems called Abell 226. The Abell group is situated at a distance of some billions of light-years away from us. Behind these objects there are even more star systems, they are less bright though, but still at even greater distances of about 9 up to 10 billion light-years. The light we see today coming from these systems has therefore travelled for 9 up to 10 billion years in order to reach us. This also means that we are looking back in time at a universe that existed 10 billion years ago.

and our individuality? Is death the absolute end of our lives, or do our lives continue beyond the boundaries of death? Is there a final heavenly (or hellish) destination past earthly existence, or is our short earthly existence the beginning, middle and end of the story? Theistic or atheistic philosophies will answer these questions in opposite ways leading to very different world perspectives which strongly affect everything we think, say and do. Even scientific disciplines such as physics and cosmology are strongly influenced, both directly and indirectly, by the dividing line between atheism and theism. As an interesting side note, it is precisely these sciences, combined with mathematics, that contain the initial answers to the question of whether the universe is governed by chaos or intelligence and thus, whether God exists or not. Given the impact this question has on our life, individually and in society, this really is the most important question that humans can ask themselves.

This book attempts to answer this question, by providing an historical overview of the subject and by critical philosophical analyses and scientific research. This book compares the scientific and philosophical arguments in favour of, or against the existence of God, compares atheism and theism, and analyses the underlying argumentation on both sides. It does this by focusing on some important changes in scientific thought, especially in the area of physics and cosmology where new and completely revolutionary discoveries have been made. These discoveries and insights reveal a universe that is infinitely complex, infinitely organized and infinitely mysterious. The level of organized complexity is so massive that this can only be explained logically by the presence of an all-pervading intelligence and an omnipresent awareness. Such an all-pervading intelligence can be called by any name and each label can be granted to it. God, of course, is the most obvious name: all-pervading intelligence and omnipresent awareness are qualifications that can only be attributed to God. The problem however is that the term God is burdened with a controversial history, created by humans. These controversies, without exception, stem from ignorance, sectarianism, fanaticism or a corrupted desire for power. The intention of this book is to demonstrate, based on objective and scientific criteria, that intelligence and consciousness are the driving forces behind the universe, regardless of the burdened history of what that implies. This burdened history is what it is, but it does not alter the reality of these new scientific insights and the philosophical consequences of these insights. Where science directed humanity towards materialism and atheism over the past 200 years, we now see a way of thinking in the opposite direction. This direction is of a spiritual nature and implies a scientific rehabilitation of God. The facts that science has revealed over the past decades confirm that a universe without God is untenable, despite desperate and sometimes exotic attempts to do so.

Whether God exists

This, therefore, is the central theme of the book, as the (sub) title indicates: 'Intelligence or chaos: the misconception of atheism.' This book discusses the scientific and philosophical arguments for and against the existence of God, atheism versus theism, and in more scientific terms, intelligence and design versus chaos and chance. Just about everyone understands what is meant with the term 'God', but on the other hand it is a concept with a wide range of interpretations. The general definitions of God, as we find them in the dictionaries are: Supreme Being, Infinite, Eternal, Omniscient, All-pervading, All-powerful, All-knowing, Self-satisfied, full of Goodness, Creator and Maintainer of the universe and the Cause of everything that exists. Many of these attributes overlap to some extent, such as omniscience, all pervasiveness and all-knowingness, or Supreme Being, all-powerful, creator and maintainer. Yet they do refer to subtle differences that relate to different areas of reality and psychology, such as existence, knowledge, power and bliss. Within these general descriptions there are important differences in nuance, for instance is God personal or impersonal, is he one with the creation or transcendental, beyond his creation, is he personally involved with the universe or does he exercise his control at a distance. In the Vedic tradition of India, God is described by three essential attributes, that summarizes many of the above qualifications quite neatly, namely Sat, Chit, Ananda. Sat refers to eternal and infinite Being, Chit refers to unlimited knowledge and all-pervasive consciousness, and Ananda refers to infinite Bliss and Goodness. All of these descriptions and attributes can in the end be reduced to two major and distinct schools of thought in defining the being and the nature of God, namely monotheism and monism.

Within each of these schools of thought there are important nuances and differences, but this book will primarily deal with these two main categories. According to monotheism there exists one divine Supreme Being endowed with personal, transcendental attributes. Monotheism also asserts that the world—the universe— is an emanation and creation of God. According to this view, both God and his emanations are eternal energies. The Christian doctrine deviates somewhat from this view, since creation is not considered to be an emanation, but as something that was created by God out of nothing. This is called 'creatio ex nihilo' by Christian theologians. Here, but also in other aspects, there are nuanced differences between the various monotheistic traditions. What the different monotheistic schools do agree on is the absolute unity of God, which is at its core both personal and transcendental. Within this unity there is, however, a multitude of diversity: first of all, between God and His energies, and accordingly, between His energies mutually. This principle of fundamental oneness that yet contains diversity is the essence of monotheism. In Christianity the unity of God is not entirely without controversy for the doctrine of the Trinity states that God is really three persons, the Father, the Son and the Holy Spirit, and not one person. Effectively, and implicitly most Christian theologians see God as fundamentally one, yet simultaneously many, or three in this instance. Despite this nuance, Christianity is generally accepted as a monotheistic religion. Quoting the words of Jesus in John 5.44: "How can you believe, when you receive glory from one another and you do not seek the glory that is from the one and only God?" Jesus was clearly of the opinion, as was official Jewish doctrine at the time, that God is one.

Monism also asserts that there is one divine Supreme Being, but that Supreme Being is essentially impersonal. The monistic God is often described as an impersonal, all-pervading, undifferentiated, and infinite state of pure energy, made of unalloyed and impersonal consciousness. According to monism,

Richard Dawkins during the launch of his campaign in 2008, where London busses were decorated with atheist slogans.

it is only this state of absolute unity that is real and the universe, with its diversity and multitude, is no more than an illusionary reflection of this divine energy. Judaism, Christianity, and Islam are considered to be monotheistic religions. Hinduism too, despite some commonly held misconceptions, is also, at its core and origin, a monotheistic creed. The philosophical core of Hinduism is founded on the Vedanta philosophy, which is of a monotheistic nature. On the other hand, Buddhism and certain movements within the Vedanta school, such as Advaita Vedanta, are monistic by nature. The famous Dutch philosopher Spinoza (1632 – 1677) was also a monist who saw the world as the expression of an underlying, all-pervading and impersonal reality. Spinoza identified this underlying reality with God. The doctrine of Spinoza was an important influence on the thinking of Albert Einstein. Einstein believed in Spinoza's image of God: '… a God that revealed Himself in the systematic harmony of the universe'. He did not believe in a God that interfered with the fate and the actions of man.

The two main movements, monotheism and monism have numerous variants such as pantheism, panentheism, polytheism, and deism. The first, pantheism is a variant of monism. According to pantheism, God only manifests himself in the universe and does not differ from the universe in every respect. Deism and panentheism are sub-divisions of monotheism. Deism is a movement that has been popular amongst Western scientists and emerged as a result of the scientific revolution in the 17th century followed by the Enlightenment in Europe and the United States during the 18th century. Deism is a form of monotheism, with this distinction that the deistic God does not interfere directly in the world, in human affairs and nature. The latter is, according to deism, governed by the laws of nature, which were ultimately created by God. Panentheism is a concept that is perhaps not quite so familiar. It means that God is transcendent and above creation and, at the same time, he is immanent and manifests himself in creation. Effectively, it is not really different from monotheism, which also acknowledges the simultaneous transcendence and immanence of God. Polytheism, the believe in many gods and goddesses, is sometimes a disguised form of monotheism. For instance in Hinduism, the pantheon of gods are effectively demi-gods and part of a divine hierarchy, whereby demi-gods are charged with ruling and managing the universe on behalf of, and in the service of the supreme God. Other traditions such as the polytheism found in ancient Egypt, Greece or Rome are truly polytheistic, whereby the different gods and goddesses are considered to be separate entities each with their own individual powers.

In the following treatment of theism and atheism, I primarily refer to the two main groups of theistic philosophies, which are monotheism and monism. For the sake of convenience, I indicate both traditions in this book as theistic. In later chapters, the differences between these two traditions will be explained further. In religions and theistic philosophies, in both monotheistic and monistic variants, God is defined as the Supreme Being, almighty, all-knowing, omnipresent, eternal and infinite: the creator and maintainer of the universe and of all life in the universe. Furthermore, God is described as loving and merciful. A theistic world view assumes that such a being, in whatever shape or form, exists. Moreover, this implies that the universe is an organic unity, governed from an intelligent and conscious centre.

The four propositions of atheism

The atheistic world view denies the existence of such a Supreme Being and denies that the universe is an organic unity governed by an intelligent centre. Apart from admitting that there are some basic, blind laws of nature, atheism claims that the universe consists of an infinite number of material particles that reside in an infinite and empty space. Since the particles are fundamentally separated by space, they are independent and therefore on a large scale governed by coincidence and chaos. Atheism also denies the existence of another reality, apart from or next to the material reality. One of the most leading advocates of this worldview is, without question, the ethologist and biologist Richard Dawkins. He even placed atheistic advertisements on London city buses. In his book 'The God Delusion' he defines atheism as follows:

"An atheist in this sense of philosophical naturalist is somebody who believes there is nothing beyond the natural, physical world, no supernatural creative intelligence lurking behind the observable universe, no soul that outlasts the body and no miracles – except in the sense of natural phenomena that we don't yet understand. If there is something that appears to lie beyond the natural world as it is now imperfectly understood, we hope eventually to understand it and embrace it within the natural."[4]

Another atheistic thinker quoted by Dawkins is Julian Baggini. He explains atheism in his book 'Atheism, A Very Short Introduction' as follows:

"What most atheists do believe is that although there is only one kind of stuff in the universe and it is physical, out of this stuff come minds, beauty, emotions, moral values – in short the full gamut of phenomena that gives richness to human life."[5]

Based on these definitions, but also based on the definitions of other atheistic thinkers, atheism is founded on four propositions or basic assumptions:

1. The universe consists of material particles that exist independent of each other and that move independent of each other within the infinite void. The total of the movements and interactions of these particles is governed by coincidence and chaos, combined with a number of simple and blind laws of nature. There is no central intelligent coordination within the universe and the universe is not an organic unity. This is also called 'pluralism'.

2. Nothing exists apart from or outside the perceptible, physical material reality or the world of matter. Matter is the only existing substance in the universe.

Proposition 1 and 2 together are also called 'materialism'.

3. Even if there were to be a beginning of the universe, the origin of the universe has to be ultimately simple. God is by definition a complex being and, therefore, he cannot be the ultimate cause. The existence of a complex being such as God would demand that he would have to have been created by something else.

4. The universe is imperfect from a human perspective. That imperfection manifests itself most clearly in the presence of pointless suffering that each living creature is faced with. This contradicts and undermines the position of God as almighty and merciful.

The first two propositions together are called 'materialism'; it holds the view that matter is the only real substance in the universe. In this view, it is also emphasized that matter may be one substance, but that this substance is split up into innumerable similar or identical particles, or pluralism as mentioned above. These particles are independent and separated from each other by empty space, but are interacting on the bases of a number of deterministic, natural laws. Independent particles randomly moving in empty space stands at the basis of what we call chaos, or 'a state of disorder', even though these particles are influenced by deterministic laws. In science and philosophy this is also referred to as deterministic chaos. While the term deterministic chaos sounds a bit contradictory, it means that the outcome of the movement and resultant interactions of two or three individual particles is predictable. However when applied to the movements of larger numbers of independent particles, their behaviour becomes random, unpredictable and chaotic. This is related to what in physics is called the n-body or 'many body' problem, and the universe, composed of an estimated 10^{80} atoms certainly conforms to this description. Materialism traditionally asserted that this combination of material particles and empty space is eternal and that they are both causeless. Given eternity it is assumed that random motion and endless interactions of particles would eventually lead to inconceivably complex yet functioning structures, purely by chance.

The third proposition makes an exception to the notion of an eternal universe, since it asserts that the universe may have a beginning, and that there is a possible cause to the universe. This proposition claims that, should the universe have a cause, then this cause must be of ultimate simplicity. This proposition is very relevant since modern cosmology assumes that the universe did have a beginning and has not always been there, nor will it most likely always be here.

The fourth proposition is the most important one, since in the end most atheistic arguments can be reduced to this proposition, or are indirectly derived from it. Consciously or unconsciously, atheists refer to the issue of the imperfect universe and the suffering in the world as the most probable reason why God could not exist. In the following chapters, the above-mentioned four propositions will be discussed in detail and refuted one by one relying on scientific evidence and logical arguments.

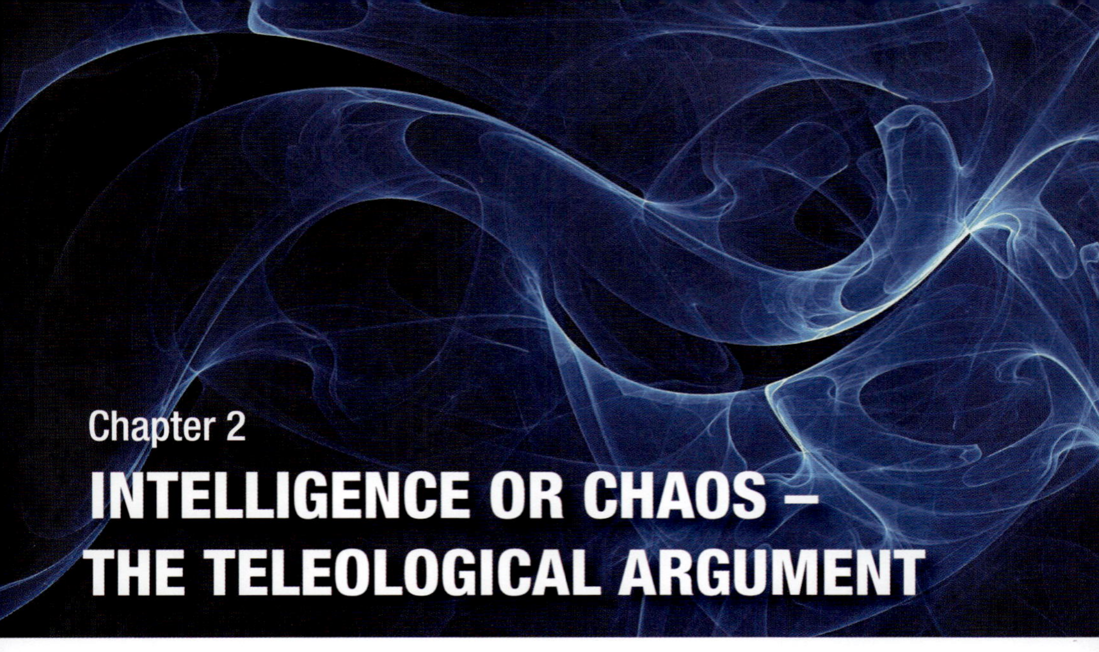

Chapter 2

INTELLIGENCE OR CHAOS – THE TELEOLOGICAL ARGUMENT

"The numerical coincidences (necessary for an anthropic universe) could be regarded as evidence of design. The delicate fine tuning in the values of the constants, necessary so that the various different branches of physics can dovetail so felicitously, might be attributed to God. It is hard to resist the impression that the present structure of the universe, apparently so sensitive to minor alterations in the numbers, has been rather carefully thought out."

Paul Davies PhD, physicist

The first atheistic proposition: complexity is the result of chance and chaos

Many committed and outspoken atheists come from the world of science and philosophy. Dawkins and Baggini for instance, are considered to be authoritative academics. They believe in the scientific method and they often consciously position themselves opposite religion— which they call 'superstition' — to show that *they* represent reason. They suggest that religion belongs to the realm of emotions and feelings, where people can vent the thought that they 'feel that there has to be something more'. They are firmly convinced that there is no, and that there cannot be, any rational or scientific foundation for the proposition that the universe arises from and is governed by an intelligent power. To quote Dawkins:

"Faith is the great cop-out, the great excuse to evade the need to think and evaluate evidence. Faith is belief in spite of, even perhaps because of, the lack of evidence."

Previous page: 3D illustration of neurons (brain cells) and nerve synapses in the human brain, the most complicated organ of the human body. The human brain consists of an average of 100 billion neurons and the human body consists of about 75 trillion cells. The complexity and the organizational level of the human body and brain are indescribable. But even the structure of the smallest atom, the hydrogen atom, appears to have a complexity and a structured balance that cannot be comprehended. From the smallest sub-atomic particle, up to the living organisms and clusters of galaxies, the universe is permeated with an indescribable level of organised complexity.

This assumption is incorrect though. First of all, the dichotomy between 'faith' and 'evidence' is a misleading notion. All knowledge is based on faith, at least to a certain degree. We know that the sun is setting this evening, and that it will rise again the next morning. No reasonable person would doubt this proposition, but strictly speaking, it is based on faith. While the sun setting and rising is based on indubitable past experiences, these experiences cannot provide a 100% guarantee that this evening and tomorrow morning the sun will continue to behave as it did in the past. All scientific knowledge is based on inductive logic, which means that the structure and sequences of past experiences are an indication of future events. The greater the amount of past experiences, the greater the certainty that they will repeat themselves in the future, but there is never a 100% certainty. As reasonable people we will not take such an extreme position of doubting every morning if the sun will be there. We have faith the sun will be there in the morning, based on reason and evidence, which is what science is all about. The key issue is whether there is scientific evidence for the notion that the universe is designed rather than a product of accidental chaos. If such evidence exists, then it is perfectly reasonable to have faith in what such evidence implies, for instance the existence of an intelligent creator.

The notion that religion is simply based on superstition, that it lacks a rational and scientific basis, and that it has no philosophical depth, is incorrect for other reasons as well. While it is of course true that most religions came into being long before the scientific revolution took place, a number of religious traditions have embraced at least the principles of the scientific method. For instance, the Vedanta tradition from India, from which Hinduism and indirectly Buddhism arose thousands of years ago, has a distinguished history of philosophical, rational and scientific inquiry into the nature of the Universe and God. According to Vedanta philosophy, there are three valid methods of obtaining knowledge, sense perception, logic and the mystic experience. The first two, sense experience and logic, are not different from the scientific method. Other religious traditions too have produced great philosophers and scientists, whose contributions to the development of human knowledge have been considerable. Most importantly, scientific developments have gone through a dramatic transformation in recent decades, a so-called paradigm shift. This transformation is ignited by an increasing number of scientific indications that the universe is unimaginably complex, and that this organized complexity is permeated by even more unprecedented organized complexity. Scientific, mathematical and logical analysis of this complexity confirm that it is impossible that this level of organized complexity could have arisen by pure coincidence and chance.

At this point, I would like to re-emphasize that although I present a particular vision with reference to the existence of God throughout this book, this vision is solely founded on scientific and rational arguments. That is in itself not new, for the presentation of rational arguments in favour of the existence of God has been done many times before. The Christian theologian and Church Father Thomas of Aquino (1225 – 1274) was one of the first Western and Christian thinkers that came up with rational arguments to prove the existence of God. Before his time, this was not considered necessary. After all, the Bible was the word of God in which God announced Himself, presented Himself, and revealed Himself. That was enough. However, as philosophy and theology developed to more sophisticated levels by the 13th and 14th centuries, it became a practical consideration to prove the existence of God with other arguments than simply Biblical Revelation. Thomas of Aquino succeeded remarkably in this regard. Even though not all of his arguments are equally strong, there is, however, one argument that still holds to this day: the teleological argument, or the argument of design. Another influential philosopher who advocated the teleological argument was William Paley (1743 – 1805), an English clergyman. His most well-known book was "Natural Theology", in which he advocated the principles of teleology, for instance using the famous Watchmaker analogy. In the course of the 19th century his book became a bestseller, even after the publication of Darwin's Origin of the Species in 1859. When Richard Dawkins wrote his

Charles Darwin

book "The Blind Watchmaker" in 1986, the title was inspired by Paley aimed at debunking Paley's arguments.

In Western philosophy the concept of teleology was first propounded by Plato and Aristotle. Teleology literally means purpose or goal. The teleological argument therefore states that many natural phenomena, including living organisms, show signs of design and purpose. Thomas stated that these natural phenomena manifest a purposefulness, cleverness and complexity that inevitably refers to an intelligent designer. Later philosophers discussed this argument in more detail, similarly referring to the complexity of nature and the complexity of living organisms, such as plants, animals and human beings. Such skilled and clever complexity, so the reasoning went, could only be explained by the presence of an intelligent creator, a designer. What is different today, compared to previous undertakings in support of the teleological argument, is the level of depth and scientific sophistication. Our scientific knowledge of the universe, of matter and of the structure of living organisms has vastly increased in comparison to the level of knowledge that existed at the time of Thomas of Aquino or of the 18[th-] and 19[th-]century scientists and thinkers. Today the teleological argument is supported by scientific evidence that was not available just thirty or forty years ago, and is therefore undergoing a major revival.

The teleological argument categorically opposes the first atheistic proposition, that the universe arose from chaos, is governed by chaos, and that its apparent symptoms of design are the result of pure coincidence. The teleological argument states the opposite: the symptoms of design are the result of design, as are the apparent symptoms of chaos. Chaos and coincidence do not exist; everything is subject to an all-pervading, infinite intelligence. The crucial question is which argument is supported by contemporary scientific insights. In this and following chapters it will be explained that recent scientific developments provide powerful and unprecedented support to the teleological argument. The logical consequence is that the alternative, chaos and chance as ultimate creators, are losing their persuasiveness and appeal.

Until Charles Darwin (1809 – 1882) published his work, the strength of the teleological argument was mostly undisputed in the theism versus atheism debate, and retained its status as a winner. Darwin suggested, however, that evolution—not design—and natural selection and coincidence were the driving forces behind creation, particularly the creation of life on earth.

Complex organs and smart functioning biological systems were the result of accidental mutations that led to an improved chance of an organism's survival. Design was just appearance whilst coincidence and evolution were reality. Evolution became a revolution. Many historians and scientists believe that no other scientific theory has changed the thought of man as dramatically as Darwin's theory of evolution. And so it seemed for a long period of time, until something changed several decades ago. New scientific insights, such as the discovery of DNA and the rise of molecular biology—but also the rise of new physics and the subsequent discovery of cosmic fine-tuning—revealed a universe of an unimaginable complexity. Scientists uncovered a complexity of unimaginable depth that not only manifested itself at a biological level in living organisms but also in the smallest structures and building blocks of the universe, such as atomic and sub-atomic particles. In the end this complexity is driven by the fundamental forces in the universe, and their values and delicate interactions have completely mystified scientists. These insights have led to the understanding that the universe must be a delicately balanced system, the so-called 'fine-tuned' universe. Fine-tuning implies that the basic forces of the universe are in a precise equilibrium, making possible the basic structures of matter, such as the very minuscule complex structures of atoms and molecules. Even minimal deviations from this equilibrium would make the existence of a hydrogen atom and the other elements of the periodic table impossible.

Another very important scientific development in relation to the teleological argument was the formulation of the big bang theory. This theory became the dominant theory regarding the origin of the universe in the 1950s and 1960s. Apart from the details and the mechanism of the big bang, the most important insight was the fact that the universe had a beginning. This insight enhanced the teleological argument since it enforced the statistical, mathematical arguments against coincidence and chaos. Coincidence and chaos were robbed of the eternity in which all improbabilities and impossibilities could become reality.

The scientific discoveries of the last 100 years have brought about a change in the direction of scientific thought. This direction used to tend toward materialism and ultimately atheism, but we currently see a clear change in direction. The teleological argument, embodied for instance in the Intelligent Design school and similar schools of thought, is making a comeback and is clearly gaining momentum. And this is not because passionate fanatics are somehow getting lots of media attention. On the contrary, the amount of media attention that promotes materialism is far greater than that for Intelligent Design. Intelligent Design is gaining strength because its arguments are impressive and convincing, especially the mathematical arguments regarding organized complexity. Even those who cannot appreciate the conclusions of these arguments have to acknowledge them, implicitly or explicitly, since mathematics is an exact science that leaves little room for interpretation. Contrary to what is often assumed, namely that a spiritual worldview belongs to the realm of feeling, intuition, and non-rational thinking, this book will demonstrate that the spiritual worldview has a completely rational foundation. Moreover, it will show that the atheistic worldview at its core, is itself based on irrational assumptions and has a dogmatic tendency towards a biased conclusion.

As stated previously, a number of very important scientific discoveries have been made in the past 100 years that have brought about this change in thinking. During the last 50 years, these developments have accelerated. The enormous technological progress in scientific research is the foundation of these developments. Advanced technology enables man to look deeper into the universe, both the micro-universe of the atom as well as into the macro-universe of infinite space. As scientists penetrate deeper and deeper into reality, it has become increasingly apparent that underneath the surface of our 'ordinary' reality there exists an extraordinary reality. This reality is unimaginably complex and organized, where the laws of time and space behave quite differently and where paradoxical contradictions exist simultaneously in mysterious ways. Mathematical analyses determining the level of complexity are impressive. Supported by the discovery of the microchip and the development of very powerful computers, mathematicians have significantly improved and accelerated their work in the area of probability and statistical analyses. The result of these analyses is baffling and is bad news for atheist materialists: Chance could not have been be the ultimate causal factor in the origin and the maintenance of the universe as proposed by the doctrine of materialism and the theory of evolution.

Below is a short summary of the most important scientific discoveries that have contributed to these insights:

1. **The discovery of the subatomic universe by Rutherford in 1911. Rutherford was the first to describe the atom as a mini solar system with a nucleus, orbited by electrons.**

2. **Albert Einstein's formulation of the special and general relativity theory in 1905 and 1916, respectively.**

3. **The discovery of nuclear energy based on the world-famous formula E = mc2, part of Einstein's special relativity theory.**

4. **The formulation of the quantum theory and quantum physics in the 1920s by Max Planck, Niels Bohr and Werner von Heisenberg.**

5. **The discovery of the expanding universe by the astronomer Edwin Hubble in 1927.**

6. **The development of the atomic bomb in 1945 by the United States of America. This development was founded on the formula E=mc2 and was preceded by research primarily by German scholars in the 1930s. Oppenheimer was the atomic scientist leading the Manhattan project that succeeded in manufacturing the first working atomic bomb in 1945. The philosophical, scientific, political and social consequences of this invention are enormous.**

7. **The discovery of the DNA molecule in the nucleus of**

living cells. This discovery revealed the unimaginable complexity of a living cell and showed it to be a very complex chemical factory. This discovery was made by James Watson and Francis Crick in 1953.

8. The formal conclusion that the big bang theory represented the most likely explanation for the origin of the universe. Hubble's discovery of the expanding universe heralded the big bang theory. The discovery of the background radiation by Wilson and Penzias in 1965 removed any remaining doubt about the validity of the theory.

9. The formulation of the anthropic principle by Brandon Carter in 1974. Carter stated that the universe is an incredibly balanced and fine-tuned system, from the smallest subatomic structures all the way up to massive star systems.

10. The formulation of the theory of the 'punctuated equilibrium' by Gould and Eldredge in 1972. According to this theory, the Darwinian model of the gradual development of the species is incorrect when one considers the fossil record.

11. The publication by Roger Penrose in 1989 in which he formulated his calculations regarding the probability of an anthropic (balanced) universe. Penrose based these calculations on the second law of thermodynamics, which is the law of entropy. Based on these calculations it became clear that it is statistically impossible that the universe came into being by coincidence and chance.

12. In general, in the 1980s many scientists started to realise that the levels of organized complexity within the universe were of massive proportions, and that the fine tuning of the universe could not have come into being just by chance. Even atheist scientists such as Fred Hoyle reconsidered their atheist points of view. The Intelligent Design School

emerged out of this development. This perception was enhanced by quantum physics and the relatively theory— the so-called new physics—revealing a universe that is energetic, mysterious, and extremely complex, and that no longer conformed to mechanical models.

These scientific developments are the most important ones from a range of discoveries that all paint the same picture of the universe: the universe is permeated by an unimaginable level of organized complexity. What is interesting is that these developments cover a broad range of scientific disciplines such as physics, biology, cosmology, and mathematics; all confirming the unprecedented level of organized complexity and the unexplainable level of a delicate balance in the universe. In this way, science has become an important tool in a rational support for the teleological argument, and therefore, in favour of the existence of God. While Thomas of Aquino lived in the dark Middle Ages, we currently have an unmeasurable amount of information at our disposal to which we have almost instant access just by turning on the computer. This information is the result of hundreds of years of scientific research, research we have been able to finance from our increasingly prosperous economies. The scientific knowledge we have gained from it has become the foundation for a number of pioneering insights. These insights are not new, but through science and technology, they can be visualized much more easily, and as such become more invasive and probing.

Unimaginable complexity

As stated above, one of the most important new scientific insights of the last 50 years is the presence of an unprecedented level of organized complexity in the universe. This complexity can be found everywhere and *at every level*; from sub-subatomic structures to biological organisms to mega-galaxies. Organized complexity is also the core of the teleological argument. Over the years many distinguished scientists and philosophers have agreed with the notion of teleology. As an illustration of this I herewith offer some quotes from a number of famous scientists. Philosopher and Jesuit Robert Spitzer wrote the following in his praised book *New Proofs for the Existence of God*:

"In the absence of a natural explanation of this extremely improbable reality, many physicists concluded that our universe is influenced by a supernatural, creative intelligence." [6]

He reached this conclusion based on calculations in which the complexity of the universe was determined.

Chandra Wickramasinghe, professor of applied mathematics and astronomy at the University College Cardiff in Wales, commenting on the inconceivable complexity of living organisms, said the following:

"The likelihood of the formation of life from inanimate matter is 1 to a number with 40,000 noughts after it ($10^{40,000}$).... It is big enough to bury Darwin and the whole theory of evolution. There was no primeval soup, neither on this planet nor any

other, and if the beginnings of life were not random, they must therefore have been the product of purposeful intelligence." [7]

The famous scientist Sir Fred Hoyle (1915-2001) remarked about these improbable numbers:

"Indeed, such a theory (that life was designed in an intelligent way) is so obvious that one wonders why it is not widely accepted as being self-evident. The reasons are psychological rather than scientific" [8]

Another observation by Fred Hoyle, who gave up his atheism based on these and comparable facts:

"A common sense interpretation of the facts suggests that a super intellect has monkeyed with physics, as well as chemistry and biology, and that there are no blind forces worth speaking about in nature. The numbers one calculates from the facts seem to me so overwhelming as to put this conclusion almost beyond question." [9]

Atheists like Dawkins acknowledge this problem, but they believe to have found a solution for this. This solution means that since each finite life span of the universe would be insufficient to let chance do its job, this finiteness should be eliminated. The multiverse theory is the perfect solution in his opinion. According to this theory (which will be discussed later in this chapter in more detail), there is not just one universe with an accompanying big bang, but an infinite number of universes and big bangs. Therefore the statistical chance that some universe out of an infinite number of universes will succeed in producing high levels of organized complexity by pure chance, has become a theoretical possibility. Cosmology professor Martin Rees comes up with the following argument in his book *Just Six Numbers*:

"If one does not accept the 'providence' argument, there is another perspective, which – though still conjectural – I find compellingly attractive. It is that our Big Bang may not have been the only one. Separate universes may have cooled down differently, ending up governed by different laws and defined by different numbers. This may not seem an 'economical' hypothesis, - indeed, nothing might seem more extravagant than invoking multiple universes – but it is a natural deduction from some (albeit speculative) theories, and opens up a new vison of our universe as just one 'atom' selected from an infinite multiverse." [10]

According to the multiverse theory, reality consists of an infinite number of universes that came into being through an infinite number of big bangs. These big bangs could be part of a series of successive big bangs and 'big crunches' that occur serially or that occur in parallel in parallel universes. They may also occur both serially and in parallel. The key point according to this theory however, is that the number of big bangs is infinite. Therefore, an infinite number of big bangs leads to the creation of an infinite number of universes and a corresponding increase in the statistical probability that at least one universe has just the right circumstances to create complex structures, including biological structures and living organisms. The adherents of the multiverse theory believe that without the limitation of the finiteness in time, the impossible becomes improbable; the improbable becomes probable; the probable becomes possible; and the possible becomes a fact. And it has happened here, for we are discussing it at this moment, as Dawkins said strikingly:

"So, the sort of lucky event that we are looking at could be so wildly improbable that the chances of its happening, somewhere in the universe, could be as low as one in a billion billion billion in any one year. If it did happen on only one planet, anywhere in the universe, that planet has to be our planet, because here we are talking about it." [11]

There is, however, one major problem regarding the multiverse theory that adherents such as Martin Rees frankly admit: it can never be empirically proven. The reason is very simple: all these parallel universes will exist fundamentally in different dimensions, with other laws of nature in another time-space continuum. We will never be able to observe them and, as Rees states, we can only conclude that they exist by means of assumptions and derivations. Therefore, such an approach completely undermines the empirically based scientific approach that atheists claim to be the foundation of their theories and arguments.

The Fine-Tuned Universe

Reality: too good to be true

Materialism states that the complexity and the appearance of design in the universe are the result of arbitrariness and chance. This arbitrariness concerns the infinite amount of basic subatomic material particles that arbitrarily connect and interact with each other within infinite empty space. This is the pluralistic universe referred to in chapter 1, which entails the notion of material particles residing in and separated by empty space. These interactions are guided by a number of basic forces, of which the four forces gravity, electromagnetism, and the strong and weak nuclear forces are the most important. These forces are expressed in the laws of nature, but physicists acknowledge that these laws are blind towards each other. It is important to note that many of the fundamental forces in nature, particularly the four forces referred to above, are constants. A constant in physics is defined as a force that exists independent of any other force, and has the value that it has, i.e. its value has no known cause. It is for this reason that balances of these values in relation to one another is so extremely mystifying, precisely because they have no cause. They are what they are, and they happened to be just right as we shall see. The forces that regulate the micro structures of the universe such as sub-atomic, atomic and molecular structures, are also referred to as the fine-structure constants. Consequently the forces that regulate the macro structures of the universe, such as planets, stars, solar systems or galaxies are called the large-scale constants.

A key element of materialism is the assumption that the chance interactions of basic material particles is an evolutionary process that takes place over vast amounts of time, whereby based on statistical probabilities, they are literally given the time and space to come up with useful structures. This point of view, namely the idea that arbitrary interactions between fundamental material particles and natural forces over a long period of time lead to strongly organized and complex structures by coincidence, is called the random universe. The supposed evolutionary process of primordial material particles is not to be confused with the Darwinian evolution of living organisms, although on a deeper level there is an important similarity. Both types of evolution refer to the emergence of new properties that were not present in their constituent components. This evolutionary vision has become a central part of the materialist version of modern science. However, in recent decades, scientists have discovered that fine tuning did not take place over billions of years of evolution and gradual development. Rather, the universe had to be fine-tuned from the very first fraction of a fraction of a second of the big bang. It was literally instantaneous.

The famous astronomer Fred Hoyle describes the big bang and the creation of basic structures of matter as follows:

"All that we see in the universe of observation and fact, as opposed to the mental state of scenario and supposition, remains unexplained. And even in its supposedly first second the universe itself is acausal. That is to say, the universe has to know in advance what it is going to be before it knows how to start itself. For in accordance with the Big Bang Theory, for instance, at a time of 10^{-43} seconds the universe has to know how many types of neutrino there are going to be at a time of 1 second. This is so in order that it starts off expanding at the right rate to fit the eventual number of neutrino types.[12]

This is a bewildering conclusion; the big bang essentially created a perfectly balanced universe in milliseconds. To try to grasp the extraordinary nature of this situation let us consider how the values of the four basic forces of nature — gravity, electromagnetism, and the weak and strong nuclear forces, are interacting, a topic which has totally mystified scientists. The values of these constants are of major importance, since scientists have determined that small changes in these values would change the entire universe. Moreover, these small changes would mean that, in almost all cases, the existence of the universe would be impossible, or that it would lead to a universe consisting solely of black holes, or a universe with one enormous black hole, or a universe without hydrogen, which would make the existence of higher forms of life and very complex organic structures impossible. In other words, the constants of the universe seem to exist in such a way and seem to be balanced in such a delicate manner that they have made the existence of the universe with its atoms, molecules, chemical elements, solar systems and star systems and, in the end, living organisms possible. This principle of the fine-tuned universe was described for the first time by astrophysicist and cosmologist Brandon Carter of Cambridge University during a lecture in Poland in 1974. He called this characteristic of the universe the anthropic principle, derived from the Greek word *anthropos*, meaning human. Carter said during this lecture that the different laws of nature seemed to be composed from the beginning in such a way that life and humanity could exist. The anthropic principle is the result of the underlying principle of cosmic fine-tuning. Cosmic fine-tuning applies to every aspect of the universe from its inception, such as time, space, primordial matter and energy, the anthropic principle zooms in to the fact the fine-tuning has enabled the appearance of intelligent life in the universe.

Cosmic fine-tuning and the subsequent anthropic principle are perfect examples of the teleological argument and in many respects define it. Fine-tuning can be considered to represent the first phase of a teleological universe; it establishes that the basic structures of the universe, such as the atomic and the subatomic particles, can exist only by the grace of the delicate balance of the four fundamental forces. Prior to the discovery of fine-tuning, it was more or less assumed that the building blocks of matter were relatively simple. Newton for instance assumed that atoms were simple, indivisible particles made of pure matter. The discovery of radiation and sub-atomic particles at the end of the 19th century shattered this notion of simple and indivisible atoms. Further research showed that atoms were complex little universes held together by fine-tuned forces. Scientists have furthermore discovered that slight changes in the values of these forces in relation to one another would lead to an unworkable mess.

Scientists distinguish between the strong anthropic principle and the weak anthropic principle. The weak anthropic principle is simply the acknowledgment that biological life-forms can only exist in very specific and balanced circumstances. Adherents to the weak anthropic principal tend towards the multiverse theory as the explanation for the very improbable circumstances that have led to our existence. On the other hand, the strong anthropic principle states that a naturalistic explanation for the anthropic circumstances of the universe are mathematically and statistically impossible. The fine-tuning of the universe demands a different explanation that will be elaborated on in the following paragraphs and chapters. In his book, *God, the Evidence*, Patrick Glynn gives a number of examples of the consequences of small changes in the values of the constants in relation to each other:

Gravity is roughly 10^{39} times weaker than electromagnetism. If gravity had been 10^{33} times weaker than electromagnetism, stars would be a billion times less massive and would burn a million time faster.

- The nuclear weak force is 10^{28} the strength of gravity. Had the weak force been slightly weaker, then all the hydrogen in the universe would have been turned into helium (making water impossible, for example)
- A stronger nuclear strong (by as little as 2 percent) would have prevented the formation of protons yielding a universe without atoms. Decreasing it by 5 percent would have given us a universe without stars.

- If the difference in mass between a proton and a neutron were not exactly as it currently is –roughly twice the mass of an electron – then all neutrons would have become protons and vice versa. Say good-bye to chemistry as we know it – and to life.
- The very nature of water – so vital to life – is something of a mystery (a point noticed by one of the forerunners of anthropic reasoning in the 19th century, Harvard biologist Lawrence Henderson) Unique among the molecules water is lighter in its solid form than liquid form: ice floats. If it did not, the oceans would freeze from the bottom up and earth would now be covered with solid ice. This property in turn is traceable to unique properties of the hydrogen atom.
- The synthesis of carbon - the vital core of all organic molecules – on a significant scale involves what scientists view as an "astonishing" coincidence in the ratio of the strong force to electromagnetism. This ratio makes it possible for carbon-12 to reach an excited state of exactly 7.65 MeV at the temperature typical of the centre of stars, which creates a resonance involving helium-4, beryllium-8 and carbon-12, allowing the necessary binding to take place during a tiny window of opportunity 10^{-17} seconds long.

The list of these so-called coincidences goes on and on and is completely mystifying. The well-known physicist Leonard Susskind writes in his book *The Cosmic Landscape* the following about this topic:

"Some (cosmologists) only see a series of remarkable consequences:

- The universe is a fine-tuned thing. It grew by expanding at an ideal rate. If the expansion had been too rapid, all of the material in the universe would have spread out and separated before it ever had a chance to condense into galaxies, stars and planets. On the other hand, if the initial expansion had not had a sufficient initial thrust, the universe would have turned around and collapsed in a big crunch much like a punctured balloon.

- The early universe was not too lumpy and not too smooth. Like the baby bear's porridge, it was just right. If the universe had started out much lumpier than it did, instead of the hydrogen and helium condensing into galaxies, it would have clumped into black holes. All matter would have fallen into these black holes and been crushed under the tremendously powerful forces deep in the black holes interiors. On the other hand, if the early universe had been too smooth, it wouldn't have clumped at all. A world of galaxies, stars and planets is not the generic product of the physical processes in the early universe; it is the rare and, for us, very fortunate, exception.
- Gravity is strong enough to hold us down on the earth's surface, yet not so strong that the extra pressure in the interior of stars would have caused them to burn out in a few million years instead of the billions of years needed for Darwinian evolution to create intelligent life.
- The microscopic Laws of Physics just happen to allow the existence of nuclei and atoms that eventually assemble themselves into the large "Tinker toy" molecules of life. Moreover, the laws are just right, so that carbon, oxygen, and other necessary elements can be "cooked" in first-generation stars and dispersed in supernovae.
- This basic setup looks almost too good to be true. Rather than following a pattern of mathematical simplicity or elegance, the laws of nature seem specifically tailored to our own existence. [13]

The main problem is that there is not a possible explanation why the laws of nature and their constants have these exact values. Scientists have tried to explain this by attempting a theory that unites the four fundamental forces into one all-embracing unified theory, but so far they have not succeeded. It is largely the mysterious nature of gravity that undermines any attempt to unify them. However, even if there was a successful unified theory, it would not explain the 'why' of this arrangement in a mechanical and causal manner. Therefore, some scientists have sought their refuge in the existence of parallel universes—in other words, in the multiverse theory. By means of this theory, they try to explain the amazing coincidence of the anthropic principle and the cosmic fine-tuning. We will discuss this 'apparent solution' in more detail at the end of this chapter.

The Statistical Probability of Chance Creating a fine-tuned universe

Jesuit and philosopher Robert J. Spitzer explores the anthropic principle in depth in his aforementioned book *New Proofs for the Existence of God*. First of all, he calculates the so-called probabilistic resources of the universe from the big bang up to the present. These probabilistic resources consist of the total number of subatomic particles in the universe multiplied by the total number of units of time in which these particles can accidentally interact with each other. Despite the colossal size of these numbers, this is a comprehensible approach to a probability based cosmic theory.

These resources are the following:

1. *The total number of protons and neutrons in the universe being 10^{80}*

2. *The total number of seconds that have passed since the big bang being 10^{16}*

3. *A maximum of 10^{43} interactions can take place per second, the so-called unit of time devised by Planck. This means an entropy per particle of 10^{43}.*

4. *The total entropy of the universe therefore amounts to 10^{123}*

Spitzer based his calculation on the work of the renowned British mathematician and physicist Roger Penrose, who carried out an analysis of all the possible universes that could have been created at the moment of creation. The term 'entropy per particle', as used by both Spitzer and Penrose in their calculations, means the number of possibilities per unit of time in which the particles can relate to each other in an arbitrary, chaotic way. Literally, entropy means disorder or chaos. Entropy and the anthropic principle are each other's enemies. Organized complexity, which is the essence of the anthropic principle, is the exact opposite of entropy. According to the second law of thermodynamics, which is the law of entropy, complex structures always tend towards simple, chaotic structures because of the process of disintegration. Entropy also implies that natural processes are basically irreversible and will only move in one direction over time. For example, when someone drops a wine glass on a stone floor, it will smash into hundreds of pieces. In this event, it is highly improbable that the pieces will spontaneously combine into a glass again. In other words, the glass is a complex structure which has to be carefully looked after in order to not disintegrate into chaos, in this instance break up into hundreds of pieces.

Within this frame of reference, entropy has everything to do with probability: complex structures are not probable—the greater the degree of complexity, the more improbable it is that the structures come into being spontaneously. Chaos and simplicity are, therefore, comparable concepts. They both refer to a situation in which objects will find the least difficult and simplest structure in relation to each other. This tendency towards simplicity and non-organized structures is called entropy and is characterized by the irreversibility of the process. Complicated structures become simple spontaneously whilst simple structures do not spontaneously become organized and complex.

At this point it is important to dwell on the concept of entropy a bit more, and to understand the underlying philosophy that determines the principle of entropy within the context of atheism and materialism. Critical to entropy is the notion that the universe fundamentally consists of two realities, particles of matter and the void. While these two realities co-exist in the same universe, they are fundamentally separated from each other since the void, in order to be truly void, is completely empty and does not possess any properties at all. Therefore the void is not able to influence the particles of matter, and thus these particles exist independently of one another. In the previous chapter this has been labelled pluralism, when defining the first of the four atheistic propositions. The material particles can only influence one another if there is direct physical contact, or when particles are influenced by a force, such as gravity. The mechanisms of these forces were not understood at the time, but it was assumed that they were extensions or some modification of matter. This physical interaction and the rules that govern these physical interactions lies at the core of Newton's laws of motion. However, Newton's laws are simple, linear and one-dimensional; every physical contact results in an opposite identical reaction. Therefore, while material particles interact, they remain independent and their motion in empty space remains random in relation to one another. These random motions create a situation whereby every combination of particles is inherently unstable. Entropy could be described as the direction towards the state where the independent particles exist in the simplest possible combination in relation to each other. This would be an eventual state of rest, whereby all material particles are evenly distributed in empty space. This would take place when the original energy that put all particles in motion in the first place, would eventually be depleted, caused for instance by frictional losses of energy in due course of time. This is indeed one of the scenarios described by cosmologists when dealing with the ultimate future of the universe, the Big Rip scenario where eventually all matter will come to a standstill in a state of absolute zero temperature.

Atheism, which fundamentally embraces pluralism, inevitably struggles with entropy. Entropy does not allow for complex and stable structures to exist perpetually, no matter how much 'chance', 'coincidence' and 'randomness' we inject into the interactions of material particles. While we do observe that entropy exists on one level in the universe, referring to the example of the wineglass above, on another level we find that the universe displays levels of organized complexity that are anthropic in nature, and anti-entropy. Countless examples can be provided of the precise correlation required between the values of the constants, such as the list of coincidences provided earlier in this chapter. All of these coincidences refer to a fine-tuned and anthropic universe. This contradiction between entropy and anthropy exists, because the philosophical and scientific notion of a pluralistic universe is fundamentally flawed and incorrect.

Based on the above data, Penrose accordingly calculates the number of possible universes that could have been created based on the given circumstances and the entropic possibilities. Since the number 10^{123} is a logarithmic function of the total number of possible universes at the moment of creation, he comes up with the improbable and unpronounceable number:

$$10^{10^{123}}$$

Robert J. Spitzer

If this number were to be partly written in figures then it would like this:

$$10^{100000000000000000000}$$

00000000000000000000000000000000000000000000000000

00000000000000000000000000000000000000000000000000

Now, if this number were to be written out fully, it would require so much space that it would fill up a large part of our known universe! This incredibly large number mirrors the improbability that of all the possibilities a universe could have emerged that possesses the exact right values of the constants required to make the existence of our universe possible. When referring to existence, I mean an existence with subatomic structures, atoms, the chemical elements of the periodic table, molecules and molecular compounds, the various fundamental forces, solar systems, star systems and ultimately the lifeforms as we know them. This number is so extremely, extremely large that Spitzer concludes:

> *"In the absence of a natural explanation for this extremely improbable reality, many physicists concluded that our universe is influenced by a supernatural, creative intelligence."* [14]

The conclusion that we can derive from these calculations is that it is simply unfeasible for the universe to have come into being by coincidence; the mathematical logic is unrelenting. Atheists will say (and I quote a random and intelligent atheist here): "Of course, there they are again with their arguments of design, complexity, statistic improbability and impossibilities. We have heard those theories over and over again." The fact is that these arguments have been presented before and they remain as powerful today as they did back then. What is new, however, is the discovery of the principle of cosmic fine-tuning, the subsequent anthropic principle and the magnitude of the mathematical consequences of these principles. Spitzer did pioneering work and the title of his book *New Proofs for the Existence of God* justifies its contents. The discovery of physical constants, especially the four fundamental forces, have given the teleological argument significant new strength. In addition to the four fundamental forces, Spitzer describes a large number of other constants, 21 in total, which all have precisely the correct values to make the anthropic universe possible. These other constants are the speed of light; Planck's constant (the smallest unit of time); various relation ratios such as those of photons and neutrons; the mass of electrons and protons related to each other; and constants that regulate the large structures and small structures of the universe, such as Hubble's constant, Boltzmann's constant, the cosmological constant, the cosmic ratio between photons and protons, the amount of free space, the electromagnetic fine-structure constant, the weak fine-structure constant, and the gravity fine-structure constant. The key and fundamental point is that each of these constants could have had any other arbitrary value, but the value they do have is exactly the right value in relation to each other and to the total, such that the creation of a fine-tuned and anthropic universe has been made possible. The fine-tuning of all these forces and constants with respect to one another is so extremely accurate that the conclusion is an inescapable fact. Chance and chaos, after a once glorious and promising career, are now relegated to the dustbin of the history of discarded and unworkable theories.

Life, the cherry on the cake

The presence of amazingly complex structures

The second phase of the teleological argument relates to the existence of amazingly complex structures, in particular those of living organisms. This is termed the second phase because the first phase relates to the fundamental structures of the universe, such as atoms, atomic and molecular compounds, energy and force fields, planets, solar systems, and star systems. The second phase relates to that which is created with the assistance of and from these fundamental structures, such as the structures of living organisms. The human body contains, for example, about 75 trillion cells, of which each cell is a chemical factory of unimaginably complex proportions. The nucleus of the cell consists of DNA, an amazingly complex molecule that is responsible for the functioning of the cell. DNA can be compared to a computer code, in which instructions are laid down that determine the functioning of the cell and the entire organism. Except that the DNA of a living cell is so complicated that it would need an encyclopaedia of a thousand volumes to write it out. If you were to lay down the DNA strings of one person in their full length, they would be as long as the distance from earth to the moon—400,000 kilometres—and then 3,000 times back and forth; a total of 2.4 billion kilometres! [15]

Another example of the unimaginable numbers that characterise the structure of cells is the number of hydrogen atoms in one human cell. One human cell contains an estimated 3,000 trillion (or 315) hydrogen atoms that are all linked and coordinated together in an unimaginable, complicated way. All this is happening just to create one living cell and to keep it alive. The human body is therefore an unimaginably complex system, and no amount of superlatives will do justice to this phenomenon. This complexity can only be expressed with the assistance of abstract mathematical and statistical representations of indescribably large numbers.

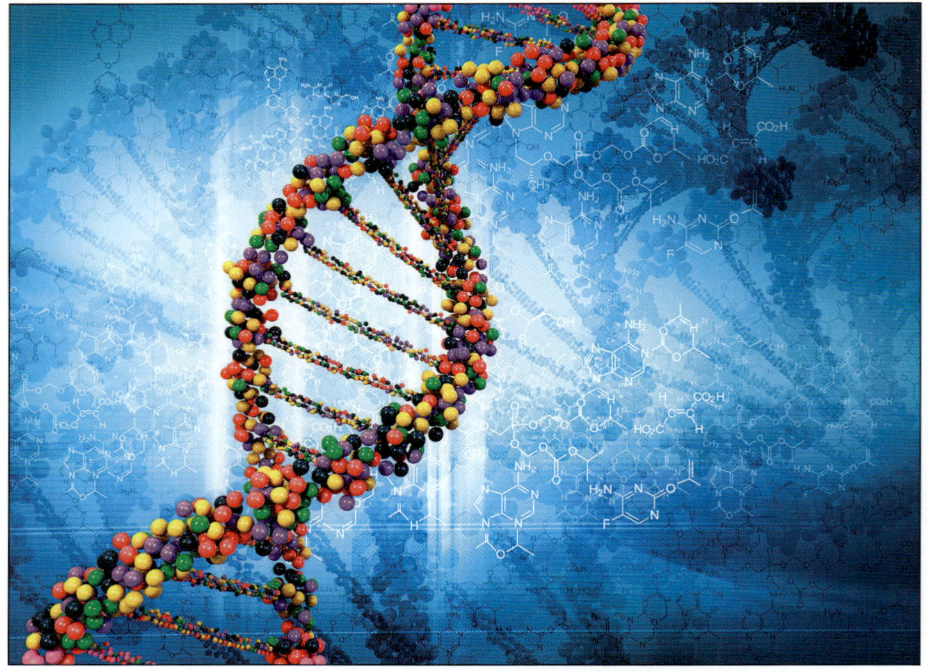

3D illustration of a DNA molecule

The common scientific explanation is still that living cells have come into being from matter coincidentally, after millions of years of evolution. This assumption —that complicated organic structures can arise from simple, non-organic structures by chance— is the cornerstone of the theory of evolution. The question, however, is whether such Complex Specified Information, or CSI, can be produced by pure coincidence, at least at the level described above. CSI is a mathematical term that describes complex structures such as machines and distinguishes it from other structures. A machine—for example, a computer—is designed and built according to a combination of many specific parts (complexity) in a specific order and structure (specification). Each machine made by man, whether it is a simple or complex machine, contains complexity and specification. However, structures that were not made by man, such as man himself, also clearly show the same characteristics of complexity and specification. William A. Dembski writes in his book *Intelligent Design* that coincidence can 'produce either complex, unspecified information or non-complex, specified information', but it cannot produce complex, specified information. The example he uses is of the typist who types an arbitrary row of letters. This generates complex, unspecified information that can look like the following (typed arbitrary by myself):

"hehdkjoihfdius dkjhhjusjsdiuyekmncxmbshsduyekjs dlksdm sdkjiuelksdlkhsdkkljsd hsduiy elk poterouweisaks dnbxcm, dlk j iuwlks adpoijfudeljksdlkjk jdyuiwpokdsnhd oosmms dojdn jhdy eiw uois dmncxhsj kdfioeiugfhjs apoxcmnskijids".

In this short, somewhat complex series of letters, nothing meaningful is written. There are, however, a number of meaningful, very short and therefore non-complex words— for example, the words 'elk' and 'pot' represent non-complex but specified information. Typing randomly cannot produce long, meaningful lines, or information that is both complex and specified. Dembski further explains:

"When the complexities become too vast and the specifications too tight, chance is eliminated and design is implicated. Just where the probabilistic cut-off is can be debated, but that there is a probabilistic cut-off beyond which chance becomes an unacceptable explanation is clear. The universe will experience heath death before random typing at a keyboard produces a Shakespearean sonnet. The French mathematician

Emile Borel proposed 10^{-50} as a universal probability boundary below which chance could definitely be precluded - that is, any specified event as improbable as this could not be attributed to chance." [16]

To make this clearer, this means that if the chance of a certain event happening is only one in 10^{50} (which is 10 with 50 zeros) or smaller, then the likelihood of this event taking place by pure coincidence is effectively an impossibility. Consequently, if this event has indeed taken place, we can conclude that it was not by coincidence.

Using Borel's probability limit as a gauge we can see that the probabilities for even the most basic elements of a living organism to have occurred by chance are so small that they are many orders of magnitude beyond the aforementioned probability boundary and therefore such a hypothesis has to be rejected. Robert Shapiro, a chemistry professor of New York University and a DNA expert, has calculated what the probability is that 2,000 types of proteins can form coincidentally in one bacteria (considering there are 200,000 different kinds of proteins in one human cell). The number that expressed this chance is $10^{40,000}$, which is 10 followed by 40,000 zeros. [17] Chandra Wickramasinghe, professor of applied mathematics and astronomy of the University College Cardiff in Wales, states the following:

"The likelihood of the formation of life from inanimate matter is 1 to a number with 40,000 noughts after it ($10^{40,000}$).... It is big enough to bury Darwin and the whole theory of evolution. There was no primeval soup, neither on this planet nor any other, and if the beginnings of life were not random, they must therefore have been the product of purposeful intelligence" [18]

The famous scientist Sir Fred Hoyle says the following about these improbable numbers:

Indeed, such a theory (that life was designed in an intelligent way) is so obvious that one wonders why it is not widely accepted as being self-evident. The reasons are psychological rather than scientific." [19]

Sir Fred Hoyle

The biologist Stephan C. Meyer says the following about the chance that life came into being by coincidence in his book Signature in the Cell:

"The complexity of the events that origin-of-life researchers need to explain exceeds the probabilistic resources of the entire universe. In other words, the universe itself does not possess the probabilistic resources necessary to render probable the origin of biological information (and therefore life) by chance alone." [20]

The mathematician Richard Thompson has calculated that the appearance of higher life-forms based on the theory of probability is impossible, with even more extreme numbers as its result:

"We can conclude that the entire course of events in S (= a system of sub-atomic particles) over a 4.5 billion year period would have to be repeated over and over again at least $10^{150.000}$ times for there to be a reasonable expectation that X (= a higher life form) would be seen in S even once". [21]

This implies that the entire history of the earth would have to be repeated over and over again at least $10^{150.000}$ times for there to be a substantial chance that a higher life form would evolve just once. Thompson's calculations and their conclusions therefore suggest an improbability that is $10^{149,950}$ higher than the point that Borel proposed. All these calculations go far beyond the probability boundary; they are impossible. This is extremely important, from both a scientific as well as a common sense point of view. Try to imagine the number we are referring to: $10^{150,000}$ times 4.5 billion is the same as 10 followed by 150,000 zeros to be multiplied by 4.5 billion.

Adherents to the theory of evolution have to some extent been able to hide behind what are unimaginable large units of time from normal human perspective, such as the fact that the earth is four billion years old. However, it is clear that these unimaginably long units of time are just a passing moment compared to the actual time required for life to have occurred by chance based on entropic limitations and statistical probability calculations for just the simplest components for life. The fundamental basic assumption of evolution has been that in the most unimaginable time period, namely eternity, everything is possible, even the most unimaginable. The problem is, however, that whilst both the earth and the universe are indeed very old, 4 billion and 13.7 billion years old respectively, that is certainly not eternity. It is important to again keep in mind in this context that the total number of atoms in the universe is not more than 10^{80}. This is an absolutely enormous number, but compared to the number $10^{150,000}$, or the possibility that higher life forms could have evolved by chance, it becomes almost insignificant. Richard Dawkins, advocate of the evolution theory, describes the impossibility of the theory of evolution as follows:

> *"So, the sort of lucky event that we are looking at could be so wildly improbable that the chances of its happening, somewhere in the universe, could be as low as one in a billion billion billion in any one year. If it did happen on only one planet, anywhere in the universe, that planet has to be our planet, because here we are talking about it."* [22]

At this point, the previously mentioned multiverse theory is not able to provide a solution at all. The emergence and evolution of living organisms has to take place in this universe and during the life time of this universe. The outcome of this process depends on the available time, the entropic limitations, the probabilistic resources, and the fundamental structures of matter that are present in this universe. And as we have seen, mathematical analyses show that these are completely inadequate to make evolution possible, in the words of Stephen Meyer:

> *"...the universe itself does not possess the probabilistic resources necessary to render probable the origin of biological information"*.

The multiverse as saving grace

A number of scientists have been trying for quite a while now to find a solution to the mathematical challenges described above; the impossibility that the universe itself with its fine-tuned constants could have come into being by chance or that the complex biological structures within the universe could have emerged by pure coincidence. To tackle this problem, a number of prominent scientists have developed the multiverse theory which was discussed briefly at the beginning of this chapter. According to the multiverse theory, reality consists of an infinite number of universes that came into being through an infinite number of big bangs. Because there are an infinite number of big bangs, this leads to the creation of an infinite number of universes and therefore an increase in the statistical chance that a particular 'lucky' universe possesses the exact right values of the constants to facilitate the creation of a working universe, the type of universe we live in. And furthermore, that it has the necessary conditions to conquer the second gigantic mathematical hindrance, which is the evolution of complex organic life from non-organic matter. The adherents of the multiverse theory believe that without the finiteness in time, the impossible becomes improbable, the improbable becomes probable, the probable possible, and the possible a fact. And it has happened here, for we are discussing it right now, as Dawkins put it so pithily.

A multiverse can come about in three different ways, either serially, parallel or both. In the serial version, several universes can come into being as the result of a cyclic process, where the big bang is followed by the big crunch. According to this theory, the big bang causes an expansion of the universe that will come to a halt after a while. Subsequently, due to gravity, an implosion will follow that will lead to a new singularity. This is the so-called big crunch. Consequently, this singularity would lead to another big bang and so on. This cycle, a kind of 'bouncing' universe, was a fairly plausible option for a while, until cosmologists discovered in 1998 that the expansion of the universe speeds up instead of slowing down. Therefore, the big rip scenario is more probable; the universe keeps expanding for ever until it basically gets ripped apart. This ripping apart will coincide with the universe freezing to death, where the temperature will reach absolute zero. Since the Big Rip does not allow for a bouncing universe, the multiverse would have to rely on the existence of parallel universes, which is the second way in which a multiverse can come about. This notion is based on the assumption that there could be an infinite number of parallel universes that exists simultaneously with and parallel to our own universe. Each of these universes could themselves go through bouncing cycles, depending on big crunch or big rip scenario's, in which case the universes in the multiverses would be both bouncing and exist in parallel realities. The multiverse hypothesis can be divided into a moderate and a radical version. The moderate version basically states that the multiverse consists of various domains, but each of these domains are still fundamentally ruled by the same laws of physics.

Robert Spitzer explains this as follows:

"There is much loose talk, even among physicists and philosophers, of "many universes." In all the theories we have been talking about, there is really just one universe, if we mean by the universe the entirety of physical reality that is in any way physically connected to the world we experience. In "multiverse" models, the universe has many "domains", but they are all parts of the same structure that is governed, ultimately, by one set of fundamental laws. Those fundamental laws may be realized in different ways in different domains, but the fundamental laws are the same in every domain, and the domains physically interact with each other in ways governed by those laws. Similarly, in bouncing or cyclic universe scenarios,

> *there may be different cycles, but all those cycles are all part of*
> *a single process governed by one set of fundamental laws."* [23]

With regard to the "bouncing" universe Spitzer remarks that it is highly unlikely that this bouncing cycle could go on forever:

> *"Is it possible that the universe has been bouncing like this*
> *forever? There are three indications that this is highly*
> *unlikely: 1) the radiation paradox 2) the entropy paradox 3)*
> *the increase in cyclic expansion. All of them are related to the*
> *second law of thermodynamics, the law of entropy. In every*
> *cycle of the bouncing universe, irreversible processes go on.*
> *For example: stars shine. They shine because they are much*
> *hotter than their surroundings, they release energy to their*
> *surroundings in the form of light. And, just as in the case*
> *of a hot cup of coffee releasing energy into its environment,*
> *this is a "thermodynamically irreversible process" - it increases*
> *the entropy of the universe. It also increases the amount of*
> *radiation filling the universe."* [24]

This is a very logical analysis of the multiverse theory that also exposes its weakness. The weakness is that in this version of the multiverse theory, the fundamental laws, including the law of entropy, have to be applicable to the other universes. But, if the laws of nature are the same across the various universes, then statistical strokes of luck are impossible.

As we can see, the moderate multiverse hypothesis does not solve the problem of the atheist, for he is still stuck with the laws of entropy that govern these universes. Therefore a number of scientists and philosophers have adopted a radical version of the multiverse. The radical version implores that parallel universes exist in a totally different reality, which is disconnected from our universe and its laws of physics. They exist in entirely different time-space continuums and they are not part of the multiverse or megaverse as some scientists prefer to call it. They actually exist in a completely different reality, with completely different laws of nature than ours. There is no connection with our physical, perceptible world, or our world of experience.

Cosmologist professor Martin Rees, gives the following explanation and description in his book *Just Six Numbers*:

""If one does not accept the 'providence' argument, there is another perspective, which – though still conjectural – I find compellingly attractive. It is that our Big Bang may not have been the only one. Separate universes may have cooled down differently, ending up governed by laws and defined by different numbers. This may not seem an 'economical' hypothesis, - indeed, nothing might seem more extravagant than invoking multiple universes – but it is a natural deduction from some (albeit speculative) theories, and opens up a new vison of our universe as just one 'atom' selected from an infinite multiverse." [25]

However, there is an enormous problem regarding this radical version of the multiverse theory that adherents such as Martin Rees admit quite frankly: it can never be empirically proven. The reason for this is simple; all other universes will exist in fundamentally different dimensions, with other laws of nature in another space-time continuum. The consequence of this is that they can never be perceptible from our point of view, and therefore, factually, they do not exist. As Rees puts it, we can only conclude that they exist by means of suspicions and inferences, which means this is a purely theoretical hypothesis that lacks any empirical and scientific foundation.

But even if for the sake of argument we allow the existence of parallel universes in accordance with the radical multiverse theory, the problem would in any case not be solved. These universes would have to be entirely different from our own universe in every respect, and therefore they could not possess any features in common with our own universe. There could be no time, space, matter, energy or forces possessing certain values. Since there are no comparable forces with comparable values, statistical probabilities based on an infinite number of tries of comparable values, simply would not exist either. The whole multiverse exercise, entirely hypothetical from the outset, fails before it even begins. Yet, the radical multiverse theory is the only real argument that atheists such as Dawkins can employ against the teleological argument. They are now left empty handed as the above arguments clearly demonstrate.

Another physicist, Paul Davies, says the following about this:

> *"... the multiverse theoreticians admit that the 'other worlds' in their theory can never—not even in principle—be observed. Travel between quantum variants is prohibited."* [26]

This is a truly ironic turn of events. In the past scientists accused religious thinkers and theologians of using fundamentally unprovable hypotheses, but we now see that some scientists are seeking refuge in fundamentally unprovable hypotheses themselves, such as the existence of the multiverse. The evidence for intelligent guidance in the universe is growing stronger, while indications for chaos and coincidence as ultimate creators has become more and more improbable and the justification for this has inevitably become more and more exotic and extreme. The tables have indeed turned. An argument often employed by atheists is that God is the product of ignorance; 'the God of the Gaps', wherein "gaps" refer to the gaps in our knowledge. In this regard, the tables have also turned. First of all, this reproach is dated: the teleological and anthropic argument indicate this. However, if one does continue this train of thought, then the multiverse theory is itself more than suitable for this qualification. The multiverse theory only consists of gaps, an infinite number of gaps, each parallel universe representing a gap. A parallel universe is after all impossible to detect and verify—a position of structural ignorance without any possibility towards improvement. The launch of the multiverse theory as the last line of defence against atheism is a fairly desperate and remarkable attempt to save chance and chaos from their inevitable demise.

Complexity and intelligence

In the final analysis, the presence of organized complexity can only really be explained by the presence of consciousness, intelligence and information. As was elaborately indicated in this chapter, chaos and coincidence are totally inadequate as possible explanations. There is no other reality or function that can be compared to the characteristics of consciousness. Our everyday experiences and scientific insights confirm this. Yet, consciousness and intelligence are conspicuously absent from atheistic explanations of the universe. Or are they?

In my many discussions with atheist thinkers, their arguments in support of atheism and in trying to explain organized complexity, boil down to a combination of the following points:

1. *disconnected particles moving in empty and infinite space over a vast period of time, preferably eternity, producing complexity by chance;*

2. *the presence and function of a number of simple, natural forces; the laws of nature,*

3. *the presence of a mysterious self-organising ability of matter.*

Of these three functions, 1 and 2 have already been discussed and refuted by the teleological argument. Item 3, self-organization, is an argument I have frequently encountered in discussions with atheistic philosophers, which argument they put forward as a way to eliminate the need for an all-pervasive intelligence. Self-organization is however a vague and misleading term, based on semantic jugglery, for self-organization is effectively identical to consciousness and intelligence. Only consciousness has the ability to carry out what we call self-organization, and in later chapters the function and characteristics of consciousness will be elaborately discussed.

The remarkable aspect of the atheist point of view is that all kinds of great characteristics are attributed to the universe: mass, energy, motion, power, attributes like sounds, colours, shapes, tastes, smells, radiation, light, time, space, all of these in an infinite manner. These characteristics are, in the end, all part of the eternal, causeless and intrinsic characteristics of matter and energy that comprise the universe. It is quite special and arbitrary that between all those amazing characteristics the most amazing and most crucial of all is not present, which is consciousness. Even the most imaginative science fiction writers of a century ago could not have dreamed of the attributes of the universe, of matter, energy, time, and space as they have presented themselves now to scientists. Nevertheless, it is claimed that the universe and its origin in the singularity and the big bang are dumb and unintelligent in the end. It is even more bewildering that this dumb and unintelligent universe has produced intelligent creatures like ourselves, all purely by coincidence and against all statistical impossibilities. It implies that these intelligent creatures have now become smarter than their unintelligent creator, the Dumb Universe. This would truly be quite remarkable and strange. The main question, however, is whether it is true of not.

Given our current knowledge the proof is overwhelming. The universe— and its cause—is not dumb, but is hyper intelligent. It is so intelligent that it is even able to fool us and make us believe that it is dumb. The scientific proof that the level of complexity in the universe cannot possibly arise from chaos and coincidence is overwhelming. This is also the most important conclusion of this chapter: the universe is governed by intelligence at all times and at every level. This intelligence is God, and it represents one of the most characteristic qualities of God. God has many characteristics, which we will discuss in later chapters. However, one of His most distinguishing and characteristic qualities are an all-embracing consciousness and an infinite intelligence, with which qualities we are all confronted constantly and intensively.

Chapter 3
MATERIALISM—MATTER IS THE ONLY REALITY

"The universe is nothing more than atoms and the void"

Democritus, Greek philosopher (460-370 BC)

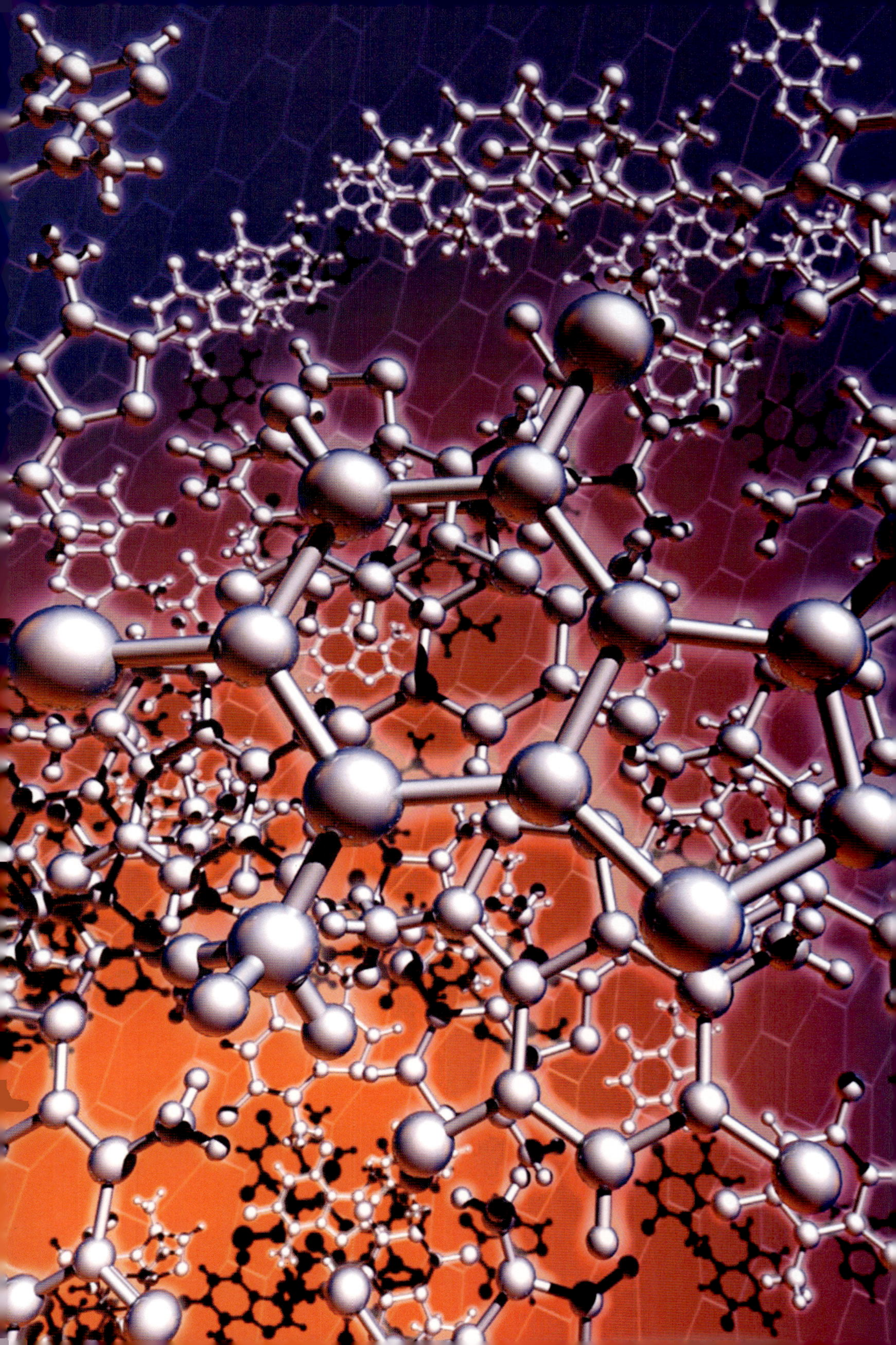

The second atheist proposition: matter is the only reality

The teleological argument is the principal and most important argument in favour of the existence of God and refutes the first atheist proposition in a convincing way. The second atheist proposition states that nothing exists outside of the material reality. This proposition is at least partially the result of the success of the scientific method, whereby only 'direct and hard observation' became increasingly the criterion of what exists and of what is real. The scientific method led to what we now call 'modern science', and no other development in the history of humanity has changed our society as much as the rise of science. We can rightly call this development a revolution, the scientific revolution. Therefore, this chapter will include a short history of science since its birth in the 15th and 16th century and its role in the creation of a new worldview: materialism, and as an extension atheism.

This worldview, just like the scientific revolution itself, has had an enormous influence on our society. Materialism and atheism have caused a significant upheaval, considering that until the 18th and 19th century many world views had a religious foundation. For this reason, in the following chapters the developments of science with materialism in its wake, will be examined in more detail. This will include a philosophical and historical perspective on how they came into being, what their basic assumptions are, but especially where materialism as a scientific theory is seriously lacking. In this respect, it is interesting to point out that whilst science has initially inspired the development of materialism, science has now become the foundation for its dismantlement.

Materialism – the modern scientific world view

The rise of science

Science arose from natural philosophy. This philosophical discipline was aimed at understanding nature and natural phenomena. Since antiquity humans were fascinated by natural phenomena—the movement of the planets, the various forces of nature, the changing of the seasons, the workings of chemistry, the mystery of living organisms—and they tried to understand them. Often times, patterns were discovered in various natural phenomena and man tried to analyse and predict them. In the Middle Ages in Western Europe, such patterns were mainly explained based on Christian theology. During the 15th and 16th century, a number of these explanations, such as the structure of our solar system, were subjected to great scrutiny. The validity of established explanations and theories was being questioned and reasonable doubts were forming, resulting in a swift increase of intellectual activities. Critical research of existing theories was combined with sensory, experimental observations of natural phenomena. The importance of systematic observation was eventually emphasized as the most important source of natural knowledge. Gaining knowledge by means of direct sense experience and observation is called empiricism and became the foundation of what we now call 'modern science'. The first philosopher to emphasize such a focus on empirical methods was Francis Bacon (1561-1626).

John Locke

The most extensive and influential explanation of empiricism was provided by John Locke (1632-1704), a British philosopher of the early Enlightenment era. John Locke stated in his book *Essay Concerning Human Understanding* (1690) that all knowledge arises from sensory experience. John Locke became the main founder of the school of British empiricism which went on to profoundly influence science, philosophy, and politics for centuries, strongly affecting various social developments at the time.

The political ideas of John Locke are also considered to be the foundation for the development of modern liberal democracy. He had a major influence on the establishment of the United States' Declaration of Independence and its Constitution, making the U.S. the first constitutional democracy in the world. Many historians believe that the famous phrase "Life, Liberty and the Pursuit of Happiness" in the new country's Declaration of Independence was written by John Locke. His liberal, political philosophy had everything to do with his metaphysical philosophy that was aimed at a pragmatic approach of reality—a reality that could be verified based on observation and common sense. In 1683, John Locke fled to the Netherlands, where he stayed until 1688 because of his suspected involvement in a political conspiracy. In the Netherlands, he had a lot of time to write, to expand his theories and to develop his *magnum opus*, the *Essay Concerning Human Understanding*, to perfection. After his return from exile, he published most of his work. John Locke is considered by many to be one of the most influential philosophers in the history of man, especially in modern history.

The emphasis on observation and sensory experience of the world and its objects is an important part of the scientific method. Observation—the ability to measure and touch objects, to see, to hear, to taste or smell them—became a crucial means to understand and determine the nature of reality. Therefore, as science developed it became focused on a reality of touchable and observational objects: the world of matter. In many ways matter was reality and reality was matter. Everything that one could not see or touch was less real or was rejected as just a thought or a fantasy. Supporters of the empiricist school rejected the notion that through ideas and reason alone one could obtain indubitable knowledge of the world and the universe. The importance of logic and mathematics was acknowledged, but they were thought to be of secondary importance and themselves an indirect result of observation. One philosopher said, 'Only from experience can one learn the truth'.

Nevertheless, scientists had fierce discussions about questions such as whether human reason (rationalism) or human observation (empiricism) were the most important ingredients for scientific research, and the acquisition of real knowledge. In the end, most scientists acknowledged that sensory observation and logic enhanced each other and therefore both were considered to be important parts of the 'scientific method'. For example, observations and experiments were used to study, measure and analyse nature while reason and mathematics were used to interpret, understand, and explain the underlying connections between the observed objects. These underlying connections could then be expressed in mathematical equations. Scientists acknowledged, however, that experience was the most important source of knowledge; reason was a tool, a derivative of experience and an addition to experience.

Materialism

Because of this emphasis on experience and observation of physical, material objects, science inevitably directed itself towards mechanistic materialism, or, in short, materialism. According to materialism, matter is the only substance in the universe. All things, including living creatures, are only expressions and manifestations of that underlying substance. Materialism, furthermore, claims that matter is indeed one substance, but divided into countless and basically identical particles, separated from each other by empty space. These expressions and manifestations are generated by the endless and complicated interactions of attraction, repulsion, and physical contact between material particles. These interactions are ruled by basic laws, such as the Newtonian laws of motion. The forces of attraction, repulsion and physical contact are predictable when they exercise their influence on the connection between two individual material objects, or limited groups of material objects. However, between the countless material objects, for example, the (almost) infinite number of atoms in the universe, the material objects interact in an arbitrary and chaotic way. According to materialism, there is no large scale coordination of the multitudes of material particles moving around in empty space. A crucial part of the materialistic world view is, therefore, the fundamental incoherent multitude of particles, or pluralism, as mentioned in the previous chapter.

The scientist and Nobel laureate Jacques Monod explains in his book *Chance and Necessity* how the universe is governed by coincidence and necessity. While individual particles are governed by mechanistic laws of attraction, repulsion and collision, there is not a coordinating cohesion between all particles and force fields. They are 'blind' towards each other and are in the end governed by chaos. While Monod was particularly referring to the chance forces that guide Darwinian evolution, his statement can, based on the doctrine of materialism, be extended to the fundamental workings of the universe. Monod expresses this as follows:

"…only coincidence and coincidence alone is the source of each innovation, of all creations in the biosphere. Pure coincidence, absolutely free, but blind is the origin of the magnificent and amazing construction of evolution …" [27]

David Hume

The famous philosopher David Hume (1711-1776) formulated this (somewhat laboriously) as follows:

"A finite number of particles is only susceptible of finite transpositions, and it must happen in an eternal duration that every possible order or position must be tried an infinite number of times….. the continual motion of matter, therefore, in less than infinite transpositions must produce this economy or order and by its very nature that order, when once established supports itself for the many ages." [28]

In short: once you take an infinite amount—or at least a massive amount of particles that will start an infinite amount of interactions within eternity—then they coincidentally will form stable combinations now and then. These stable combinations will perpetuate themselves and form the basis for the complexity that we can witness in the universe. A famous and hilarious example that illustrates this worldview is the one of the typing monkeys. This goes as follows: take an infinite number of monkeys and let them type forever on an infinite number of typewriters and in the end they will produce, purely by coincidence, the works of Shakespeare.

The analogy of the monkeys is attributed to the mathematician Emile Borel, who was quoted before within the scope of statistical probability analyses. He used this example in an article that he wrote about this topic in 1913. Mathematicians agree, by the way, on the hard-working monkeys: even within the life span of a universe of tens of thousands of billions of years, the typing (by that time, no doubt, very tired) monkeys cannot even come close to such a feat. Even if we filled the entire universe with monkeys, where each monkey was the size of an atom, then they would still never be able to produce the works of Shakespeare. Computer simulations with virtual monkeys that work in a time span thousands of times faster than ours, have not even produced one page of the works of Shakespeare in millions of years. A larger group of computerized monkeys needed 2,737,850 million billion billion billion years to produce the word combination: 'Rumour, open your ears' that occur in *Henry IV, Part II.* [29] The following phrase was '9r"5j5&?OWTY Z0d'. So, not really an entertaining play.

The essence of materialism is that the appearance of complexity and design in the universe can be attributed to pure coincidence and chance. Chance makes God superfluous and leads to the aforementioned Random Universe, a universe governed by blind chance and necessity. The essence of materialism is further that there is only one substance in the universe, and that is matter. Nothing exists outside of the material reality, no higher or spiritual energy, no supernatural forces, no mysterious guiding intelligence and no God. Materialism, as defined on the bases of these two principles, leads inevitably to an atheistic worldview. And since materialism has become the dominant scientific worldview in the 20th century, it is not surprising that atheism has been strongly on the rise for quite some time now.

Attractiveness of materialism

Materialism has an additional attraction: it provides a well-organized worldview, with an apparently clear vision of the universe. It appears to correspond to our observation, our intuition, and our common sense. We feel, see, smell, hear, and taste a world of material objects. We experience this world in daily life and we use this world and its objects in a very practical manner to provide for our livelihood and to survive. The description of the world according to materialism corresponds in many ways to our everyday perception.

Furthermore, the observations and descriptions of the universe that were made by scientists after Newton and during 18th and 19th century confirmed our everyday observations. We see an object moving in space from A to B, and we see time passing from past to present and then toward the future. We experience time and space as separate entities and material objects as massive objects, just like the great physicist Sir Isaac

Newton (1642-1727) described it. Newton's laws of motion, and their related technological applications work perfectly well in practice. We created the Industrial Revolution with them, and this led to prosperous economies where millions of people earn their daily livelihood. This is the daily reality that we know all too well and that perfectly matches the materialistic worldview.

However, appearances can be deceiving and our common sense can mislead us. There are many examples of this type of deception. Our common sense makes us believe that the earth is flat, and up to about 500 years ago, many people actually thought that to be the case. When we walk down the street, no matter where we are, the earth, indeed, seems flat. Our sensory perception is too limited and the earth is too big for us to observe its curvature. Only after humanity studied the matter in more depth did it become apparent that we are living on a globe and the Flat Earth theory appeared to be wrong. Travelling around the globe in airplanes, space travel and the shape of other observable celestial bodies, painted a very different picture of planet earth then our immediate sensory experiences. Let us take another example: from the perspective of earth, it seems that the days and nights are created by the sun revolving around earth. Again, our senses deceive us. Earth rotates on its axis, and, apart from that, earth revolves around the sun in 365 days. Until the writings of Copernicus and Galileo in the 15th and 16th centuries, humans really thought that earth was the centre of our solar system and that the sun rotated around the earth.

Time and time again it is apparent that our observations are not necessarily incorrect, but that they are incomplete. Since our 'common sense' is mainly based on our immediate observations, our common sense can therefore be misleading. Only after we gather additional information, based on further and more profound observation and analyses, is our common sense able to complete and correct our incomplete observation. Except for a few fanatics, modern man in the year 2017 will not think for a moment that the sun rotates around the earth or that the earth is flat. He has processed the additional and corrective information into his common-sense vision of reality. However, there is something odd about this corrective ability. Scientific developments that started in the beginning and the middle of the 20th century—in particular Einstein's theory of relativity, quantum physics, the big bang cosmology and the anthropic principle—should also have such a corrective effect on our materialistic worldview. These new scientific insights, as will be further discussed in depth in the following chapters, show that the materialistic world view is incomplete and is also no longer correct. On the one hand, such corrective developments form a normal pattern in the growth of knowledge and science. The problem is, however, that even though new scientific insights contradict materialism, science does not (yet) acknowledge this explicitly. Scientists have not yet sufficiently digested these new insights, so they are not able to put them into an understandable framework. Therefore, they revert back to the old paradigm of materialism.

Indeed, the materialistic worldview is still taught in schools and universities, despite its shortcomings. It is the main scientific view of the world and of the universe and we are all taught this particular view. When we say science, we are actually saying materialism, and perhaps more nuanced, atheism. There are two important causes for this. On the one hand, accepting these shortcomings inevitably leads to a theistic worldview. Scientists still resist this possibility, however, as there is still a lot of prejudice. Scientists believe that they have finally moved past the theistic worldview after a lot of effort and struggle that lasted centuries. Moreover, many theistic worldviews revert to fundamentalist basic assumptions and have often not been explained and clarified in scientific and philosophical terms, which makes a scientific discussion difficult or impossible. These deficiencies should be overcome, as will be discussed in later chapters.

Three important scientific developments

Since the beginning of the scientific revolution about 400 years ago, scientists have made an impressive series of scientific discoveries. They have discovered the structure of our solar system, the laws of motion and gravity, the structure of chemical elements, the inner workings of the atom, the laws of electromagnetism, and they have discovered nuclear energy and the microchip, just to mention a few. All these developments and discoveries have been phenomenal.

Within this historic arsenal of scientific feats, there are three scientific developments in particular that have become the foundation for materialism. These are the discovery of the laws of mechanics by Newton in 1687, the discovery of the laws of chemistry and the periodic system of Mendeleev in 1869, and the formulation of the theory of evolution by Charles Darwin. Indeed, particularly Darwin's theory has been considered in many ways as the 'completion' of materialism. His doctrine reduced biological organisms into material processes; he apparently eliminated the teleological argument and he made God superfluous. Darwin's theory of evolution was introduced in 1859 with the publication of his book *On the Origin of Species*. These three scientific developments together have had a crucial influence on the establishment of materialism and the rise of atheism.

The laws of mechanics: Newton

The discovery of the laws of motion and mechanics by Isaac Newton was the first of these three important developments.

Scientists had long been interested in observing the movement of material objects and the mutual exchange of forces between them. It had been suspected and assumed that these processes took place based on laws that were governed by mathematical and logical principles. These laws were discovered and elaborated on by Newton in his famous book *Philosophiæ Naturalis Principia Mathematica*; in short, the Principia that he published in 1687. These laws were later named after him.

Sir Isaac Newton

Newton defined his now famous three laws of motion. The first law states that an object remains at rest or will continue to move unless it is acted upon by a force. The second law states that the action of a force on an object will cause the object to accelerate. The third law is the law of cause and effect and states that each object exerting a force on another object will create a force equal in magnitude but opposite in direction. Newton assumed that these laws were universally applicable and that they influenced matter everywhere in the universe in the same way. In addition to these three laws, Newton also discovered and described the universal laws of gravity, a phenomenon that he found quite difficult to grasp and that he could not explain. Nonetheless he succeeded in quantifying gravity and captured its workings in mathematical equations. By describing the universe as a machine-like structure, governed by universal laws that could be expressed in mathematical equations, Newton enabled materialism to make a major leap in its development. However, Newton himself was not a materialist in the restricted sense of the word. He was a deist who regarded God as the creator of the universe, and the creator of the laws of nature through which he exercised indirect control over the universe.

Materialism is, by the way, not a new theory. The Greek philosopher Democritus (460-370 BC) was an adherent of materialism. He regarded the universe as 'nothing but atoms and the void.' The Indian philosopher Carvaka is considered to be the founder of a school that propagated materialism and atheism about 2,500 years ago in India. According to Carvaka, the universe existed only of matter, earth, water, air, and fire, and therefore wise men should enjoy the temporary earthly pleasures for as long as possible. Newton lifted materialism, however, to a higher level since he created a logical and mathematical framework in which natural phenomena could be explained. Up to then, they had been attributed to supernatural forces or they were just not understood. Newton made the universe understandable, orderly, and conveniently arranged. Furthermore, the laws of Newton confirmed what we could already perceive based on our everyday experience and common sense: matter has mass, objects move and are heavy, they influence each other directly, and this influence is predictable.

As previously stated, science has been of major importance regarding the creation of the 'mechanistic materialism' worldview. The 'mechanism' of these mechanical laws consists of three principles: physical contact, attraction, and repulsion. These principles determine all known forces and movements of the universe. Physical contact is best known and arises from one massive object touching another massive object, which therefore influences its movement. This principle of physical contact has been the cornerstone of Newton's discoveries. The mechanism was understandable, since it was physical and therefore perceptible. Newton also overserved that the movement of objects can be influenced at a distance, apparently without physical contact: we are all aware of the famous example of Newton watching an apple falling from a tree which led to the formulation of the law of gravity, or universal attraction. Newton did not understand the underlying mechanism of gravity but nevertheless, he accepted its existence. He did however succeed in describing and quantifying gravity with mathematical equations. The force of repulsion occurs when particles come into contact with identical electromagnetic charges. In Newton's day, however, these forces were not yet discovered or understood. The leading theory was that the forces of attraction and repulsion—and in general the influence of objects at a distance—occurred by means of some form of physical contact. Therefore, scientists had proposed the existence of an invisible substance that penetrated the entire universe: ether. Just like air facilitates soundwaves, ether facilitated more subtle waves, such as light, magnetism, and even gravity. Later research, however, indicated that ether did not exist. Light and gravity appeared to be part of force fields that existed outside and separate from material particles. At that time, this was difficult for scientists and philosophers to grasp, for this went against intuition and the mechanistic foundation of scientific thought. Indeed, as mentioned before, even Newton found it difficult to actually understand and explain the functioning of gravity.

The laws of chemistry: the periodic table

The second major scientific development was the discovery of the periodic table of chemical elements. The periodic table is a table with an overview of the series of chemical elements based on their atomic structure. It shows the progression of the smallest element, the hydrogen atom, up to the heaviest elements. We have all been taught the periodic table during chemistry class in secondary school and this table became an important pillar of modern science. In the 19th century, scientists discovered that the chemical characteristics of matter came into being through alterations to the internal structure of the atom. By the 'simple' addition of an electron and a proton to one chemical element, a totally different chemical element was created with entirely new characteristics.

In turn, this combination of various atomic elements led to the formation of molecules and the vast, near infinite variety of molecular compounds and chemical substances. The periodic table was formulated for the first time in 1869 by the Russian chemist Dmitri Mendeleev. Mendeleev was not aware of the internal structure of the atom at the time, but nevertheless his version of the periodic table was surprisingly accurate. By the end of the 19th century the structure of the atom was exposed in more detail. Scholars discovered that the atom consisted of a nucleus with a positively charged particle, the proton, and a neutrally charged neutron. The nucleus was found to be encircled by negatively charged electrons.

The periodic table starts with the smallest element, hydrogen. Hydrogen, 'H' in chemical jargon, consists of a single electron circling around a nucleus with just 1 proton and sometimes a neutron. The following element in the periodic table is Helium, consisting of 2 electrons, with 2 protons and neutrons in the nucleus. Every time an electron is added, an entirely new element comes into being, with entirely new chemical characteristics. There are currently 118 elements known to us, with the last element being ununoctium with 118 electrons and 118 protons. Small mutations in the structure of an atom, an addition of an electron or an imbalance in the number of protons and neutrons can lead to dramatic differences in the characteristics and behaviour of an element. The difference between gold and lead is only 3 electrons. Gold has position 79 in the periodic table and it has 79 electrons and lead has position 82 in the period table and has only 82 electrons. The atomic structure of both metals hardly differs, but their characteristics obviously differ greatly.

THE PERIODIC TABLE

( ) = ESTIMATES

SYMBOL — ATOMIC NUMBER — ATOMIC WEIGHT — NAME (example: H, 1, 1.008, Hydrogen)

Period	1 IA	2 IIA	3 IIIB	4 IVB	5 VB	6 VIB	7 VIIB	8	9	10	11 IB	12 IIB	13 IIIA	14 IVA	15 VA	16 VIA	17 VIIA	18 VIIIA
1	H 1 1.008 Hydrogen																	He 2 4.00 Helium
2	Li 3 6.94 Lithium	Be 4 9.01 Beryllium											B 5 10.81 Boron	C 6 12.01 Carbon	N 7 14.01 Nitrogen	O 8 16.00 Oxygen	F 9 19.00 Fluorine	Ne 10 20.18 Neon
3	Na 11 22.99 Sodium	Mg 12 24.31 Magnesium											Al 13 26.98 Aluminum	Si 14 28.09 Silicon	P 15 30.97 Phosphorus	S 16 32.07 Sulfur	Cl 17 35.45 Chlorine	Ar 18 39.95 Argon
4	K 19 39.10 Potassium	Ca 20 40.08 Calcium	Sc 21 44.96 Scandium	Ti 22 47.88 Titanium	V 23 50.94 Vanadium	Cr 24 52.00 Chromium	Mn 25 54.94 Manganese	Fe 26 55.85 Iron	Co 27 58.93 Cobalt	Ni 28 58.69 Nickel	Cu 29 63.55 Copper	Zn 30 65.39 Zinc	Ga 31 69.72 Gallium	Ge 32 72.61 Germanium	As 33 74.92 Arsenic	Se 34 78.96 Selenium	Br 35 79.90 Bromine	Kr 36 83.80 Krypton
5	Rb 37 85.47 Rubidium	Sr 38 87.62 Strontium	Y 39 88.91 Yttrium	Zr 40 91.22 Zirconium	Nb 41 92.91 Niobium	Mo 42 95.94 Molybdenum	Tc 43 (97.9) Technetium	Ru 44 101.07 Ruthenium	Rh 45 102.91 Rhodium	Pd 46 106.42 Palladium	Ag 47 107.87 Silver	Cd 48 112.41 Cadmium	In 49 114.82 Indium	Sn 50 118.71 Tin	Sb 51 121.76 Antimony	Te 52 127.60 Tellurium	I 53 126.90 Iodine	Xe 54 131.29 Xenon
6	Cs 55 132.91 Cesium	Ba 56 137.33 Barium	La 57 138.91 Lanthanum	Hf 72 178.49 Hafnium	Ta 73 180.95 Tantalum	W 74 183.85 Tungsten	Re 75 186.21 Rhenium	Os 76 190.2 Osmium	Ir 77 192.22 Iridium	Pt 78 195.08 Platinum	Au 79 196.97 Gold	Hg 80 200.59 Mercury	Tl 81 204.38 Thallium	Pb 82 207.2 Lead	Bi 83 208.98 Bismuth	Po 84 (209) Polonium	At 85 (210) Astatine	Rn 86 (222) Radon
7	Fr 87 223.02 Francium	Ra 88 226.03 Radium	Ac 89 227.03 Actinium	Rf 104 (261) Rutherfordium	Db 105 (262) Dubnium	Sg 106 (263) Seaborgium	Bh 107 (262) Bohrium	Hs 108 (265) Hassium	Mt 109 (266) Meitnerium	Unnamed Discovery 110 Nov. 1994	Unnamed Discovery 111 Nov. 1994	Unnamed Discovery 112 1996		Unnamed Discovery 114 1999		Unnamed Discovery 116 1999		Unnamed Discovery 118 1999

ALKALI METALS (group 1) — ALKALI EARTH METALS (group 2) — HALOGENS (group 17) — NOBLE GASES (group 18)

LANTHANIDES

Ce 58 140.12 Cerium	Pr 59 140.91 Praseodymium	Nd 60 144.24 Neodymium	Pm 61 (145) Promethium	Sm 62 150.36 Samarium	Eu 63 152.97 Europium	Gd 64 157.25 Gadolinium	Tb 65 158.93 Terbium	Dy 66 162.50 Dysprosium	Ho 67 164.93 Holmium	Er 68 167.26 Erbium	Tm 69 168.93 Thulium	Yb 70 173.04 Ytterbium	Lu 71 174.97 Lutetium

ACTINIDES

Th 90 232.04 Thorium	Pa 91 231.04 Protactinium	U 92 238.03 Uranium	Np 93 237.05 Neptunium	Pu 94 (240) Plutonium	Am 95 243.06 Americium	Cm 96 (247) Curium	Bk 97 (248) Berkelium	Cf 98 (251) Californium	Es 99 252.08 Einsteinium	Fm 100 257.10 Fermium	Md 101 (257) Mendelevium	No 102 259.10 Nobelium	Lr 103 262.11 Lawrencium

The discovery of the periodic table of chemical elements was of crucial importance for the development of the materialistic worldview. It showed that the enormous diversity of chemicals and their characteristics, colours, smells, forms, and structures were simply created by the addition or the removal of electrons, neutrons, and protons within the atom. The interesting thing is that each basic particle—for example, an electron—is basically identical to all other electrons. This also holds for neutrons and protons. The diversity of characteristics comes into being by their numbers, their mutual relations, and the structures in which they are placed. The symmetry and simplicity of this system cannot escape anyone's attention and it matches beautifully with the laws of mechanics. After all, the internal structures of the atom are partly determined by the attraction and repulsion of the various electrically charged particles, such as the electrons and protons, and by the weak and strong nuclear forces in the nucleus of the atom.

The discovery of the periodic table and the structure of the chemical elements also made it possible to explain the evolution and the origin of these elements from the big bang to today, through a process called 'nucleosynthesis". In general, cosmologists assume that matter that came into being immediately after the big bang, consisted mainly of the lightest element hydrogen and a small quantity of helium. Since certain parts of the universe had larger concentrations of hydrogen than others, it led to the formation of hydrogen clusters. These clusters became more and more compressed because of gravity and therefore the temperature increased. At very high temperatures of millions of degrees, hydrogen atoms will fuse together and form helium atoms, creating a nuclear reaction. This leads to the creation of a star, comparable to our own sun, in which these nuclear reactions are taking place in a massive continuous chain reaction. These nuclear reactions generate an enormous amount of energy, just like the sun. This energy is generated because the nucleus of a helium atom is just a bit lighter, meaning that it has a little less mass than the two separate hydrogen atoms. In other words, during the nuclear fusion a small part of mass is transformed into energy in accordance with Einstein's formula $E = mc^2$. According to this formula, a small amount of mass creates an enormous amount of energy. This explains why, for example, our sun—having a mass that is 330,000 times bigger than earth—can produce enormous amounts of energy for billions of years.

Stars create new and heavier elements because of nuclear fusion. They mostly create helium but they also create the elements of the periodic table following helium such as lithium, beryllium, carbon, nitrogen, and oxygen. These heavier elements also take part in the nuclear fusion process, which in turn also produces considerable amounts of energy. As was stated, this process occurs because of the enormously high temperatures that are being created by permanent nuclear reactions within a star combined with the gravity caused by its mass. Within the life span of a star this process of nuclear fusion continues up to and includes the creation of iron and nickel, numbers 26 and 28 in the periodic table. After nickel, nuclear fusion cannot produce any more energy since the mass of the nucleus of a zinc atom—the element two places after nickel in the periodic table—is heavier than the atomic nuclei that it originates from. In other words, the creation of a zinc atom from, for example, iron and helium will use up energy instead of producing energy. In this way during the life span of a star, a whole series of chemical elements is created up to and including iron and nickel.

Finally, when all the nuclear fuel—such as hydrogen, helium, nitrogen, carbon, silicon, etc.— has burned up as the star reaches the end of its life span, then the star will implode under the weight of its own mass. This leads to the creation of a supernova, in which the inner part of the star implodes into a so-called neutron star, or, if the mass of the star is big enough, into a black hole. The outer layers of the star become part of a gigantic explosion, from which enormous amounts of neutrons come into being, and which, in a split second, creates chemical elements that are heavier than iron and nickel. This process is called 'Supernova nucleosynthesis', the process through which the heavier elements in the periodic table are created. Indeed, Earth also originated from former supernova explosions, which explains why the entire spectrum of lighter and heavier elements is present on Earth.

In summary, the discovery of the periodic table and the structure of chemical elements has been of crucial importance in the development and formulation of the materialistic, mechanistic worldview. The enormous variety of qualities that we find in the vast number of chemical compounds, could now be reduced to simple building blocks and quantitative arrangements of identical material particles. Material particles that behave in accordance with the mechanical laws of attraction, repulsion and collision, thereby confirming the mechanistic nature of the universe. This ability to reduce complex arrangements to simple components, is called reductionism, and in chemistry this works particularly well. Reductionism is also the core of materialism, for according to materialism the complexities of the universe can all be reduced to rather simple material components. The laws of chemistry seem to confirm the reductionist point of view. The same applies to Newton's laws of motion. According to these laws the movement of objects and their collisions, as well as the interaction of some basic forces can all be reduced to a number of fairly simple equations. Although Newton's laws are more applicable to large scale structures and interactions, they provide an excellent explanation for the scales, speeds and changes of everyday life. These laws too, enhance the notion of reductionism, and thereby support the doctrine of materialism. A doctrine that states that only matter exists, and that the complexity of matter is the result of chance and chaos.

In the next chapter the third important scientific development that has paved the way for materialism will be discussed, Darwin's theory of evolution. The origins and tenets of this theory, as well as its significant influence on science, philosophy and society will be reviewed in a fair amount of detail. In the chapter thereafter the defects and limitations of materialism will be discussed and elaborated upon. Particularly the notion that no other substance exists outside of matter will be scrutinized and will be shown to be mistaken, based on the scientific discoveries of the 20th and 21st century. Historians have noted that the 19th century, particularly the second half of the 19th century, has been the "golden age" of materialism and atheism. Discoveries of the 20th century have turned this golden age upside down, and have reversed this trend in a significant way.

Chapter 4

DARWIN'S THEORY OF EVOLUTION

"Personally, I have little doubt that scientific historians of the future will find it mysterious that a theory which could be seen to be unworkable, came to be so widely believed." [30]

Sir Fred Hoyle, British astronomer and cosmologist on Darwin's theory of evolution in his book *The Intelligent Universe*

Charles Darwin

The importance of the theory of evolution

The third important scientific development is Darwin's theory of evolution. Considering the enormous importance of this doctrine, I have dedicated a separate chapter to this theory. Indeed, no other scientific development has had such an enormous influence on society and the world view of mankind, as Charles Darwin's theory of evolution. I herewith quote Richard Dawkins, a big and fervent adherent of Darwin:

"An atheist before Darwin could have said, following Hume: 'I have no explanation for complex biological design. All I know is that God is not a good explanation, so we must wait and hope that somebody comes up with a better one.' I cannot help feeling that such a position, even though logically sound, would have left one feeling pretty unsatisfied , and that although atheism might have been logically tenable before Darwin, Darwin made it possible to be an intellectually fulfilled atheist." [31]

Dawkins confirms in this statement that he regards Darwinism as the completion of the athllectual foundation that removes all possible doubt and closes the intellectual holes in the materialistic theory, at least according to Dawkins. While the laws of Newton and the laws of chemistry relate mainly to 'dead' matter, the theory of evolution primarily focuses on living organisms and in the end on humans. It professes to give an explanation as to how we came into being and what our place in the universe is. This has very important personal consequences, and it also directly influences our moral values as will be explained further on.

The scientist Ernst Mayr formulated this as follows:

"The relativity theory of Einstein, or Heisenberg's theory concerning statistical predictions, could hardly have had any effect on someone's personal beliefs. The Copernican revolution and the world view of Newton required some revision of traditional beliefs. None of these physical theories, however, raised as many new questions concerning religion and ethics as did Darwin's theory of evolution through natural selection." [32]

The philosopher Daniel Dennett formulated the influence of Darwinism in this unique way:

"Darwinism is a Universal Acid that eats through just about every traditional concept, and leaves in its wake a revolutionized world view, with most of the old landmarks still recognizable, but transformed in fundamental ways." [33]

The above-mentioned quotes give an indication of just how much influence the theory of evolution has on our lives. It essentially touches on almost every aspect of society, our ideas about existence and life, and our position in the cosmos.

The core of Darwinism

Charles Robert Darwin (1809-1892) developed his theory while he was travelling around the world between 1831 and 1836 on his ship, the *Beagle*. The purpose of his journey was to carry out natural research in various parts of the world as well as study geological structures and the fossil record. Darwin was very aware of the ruling theories and worldviews of his time. He had studied mathematics, physics, and theology, and his father would have loved for him to have become a priest. However, Darwin was more interested in understanding the natural world, and understanding the origin of the massive diversity of biological life forms. He was particularly inspired by the works of the geologist Charles Lyell, who had developed a theory that the formation of the Earth had taken place gradually. Because of this gradual development, it was possible to use the fossil record (amongst other things) to look back in time. Since the Earth had formed layer by layer, one should be able to find the preserved remains, or fossils, of plants, animals and humans from earlier ages in deeper layers of the Earth. The study of fossils and the fossil record is called the science of palaeontology, and it played a very important part in the development of Darwin's theories.

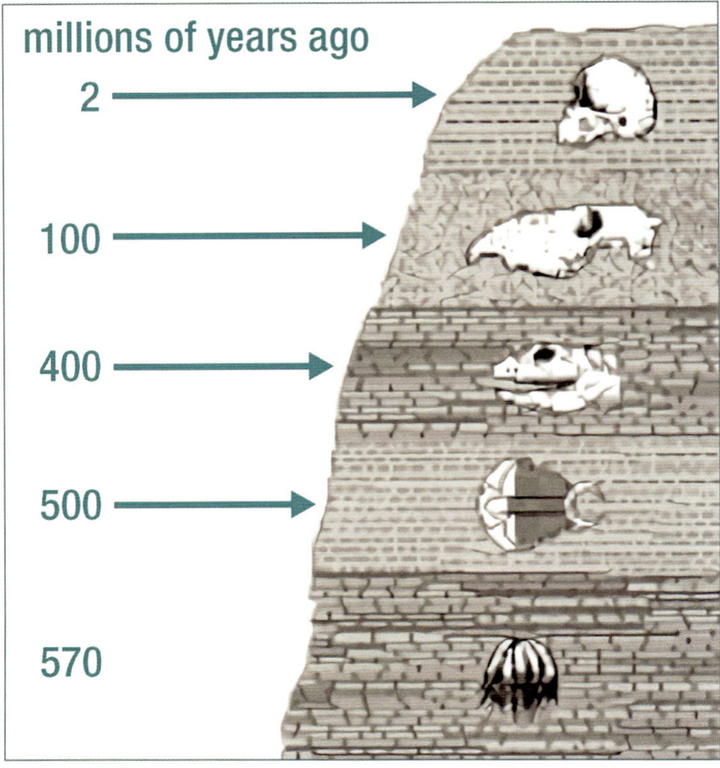

The fossil record originated because of the layered construction in the development of the earth.

The theories of Lyell had a major influence on Darwin, and the development of his theory of evolution. Darwin was also strongly influenced by some evolution-like theories that are now considered precursors to Darwin's own theory of evolution. The most well-known of these theories was developed by Jean Baptiste LaMarck (1744-1829). LaMarck's theory was very similar to Darwin's except that LaMarck believed that evolution was a guided process, while Darwin introduced the element of random chance and the process of natural selection. Another important feature of LaMarck's theory, which Darwin did agree with, stated that the acquisition of certain characteristics during the life of an organism could subsequently be passed on to future generations through heredity. A son of an athlete—for example, a muscled gymnast—would also automatically inherit the developed muscles of his father which had been acquired through training and competition. This principle is called inheritance of acquired characteristics. Darwin adopted this idea from LaMarck and it became one of the cornerstones of his theory. However, this version of heredity was later determined to be incorrect. The work of Gregor Mendel (1822-1884) regarding heredity determined that acquired characteristics were not assimilated into hereditary material. Mendel, who is sometimes referred to as the father of genetics, was an Augustine monk from Austria who had a great interest in biology. Mendel did a lot of his research in the courtyard of the monastery where he lived. Mendel's work (as well as work by other researchers on the topic) showed that hereditary characteristics are recorded in genetic material, the genes, and that even though genes can mutate, acquired characteristics such as developed muscles or acquired musical skills do not influence genetic material. Darwin, however, was unaware of Mendel's work even though they were contemporaries. After years of research and reflection Darwin finally published his book *On the Origin of Species* in 1859 in which he explained his theory of evolution. By publishing this book, Darwin became one of the most famous people in modern history and laid the foundation for a revolutionary upheaval within the scientific, philosophical, and theological communities, as well as for the population of the world at large.

Nowadays most people have heard of Darwin's theory of evolution and have some notion of what it actually means. In short, according to this theory, living organisms developed from simple unicellular organisms into complex multicellular creatures over millions of years. Through this process of development many different species have come into being, each with their own unique characteristics. And as a species develops further, it will become increasingly distinct from other species. For example, while unicellular bacteria hardly differ from one another, the difference between a human and a giraffe is huge. Darwin's theory further proposes that the development of simple creatures into complex creatures is driven by the need to survive and by the need to better adapt to a basically hostile environment. This need to survive forces organisms to adapt and to change. These changes are passed on to their offspring which in turn may change even further as they are also forced to adapt to survive. In this way over millions of years, hundreds of thousands of life forms came into being—in other words, the species as we know them on earth today.

The core of Darwin's theory can therefore be summarized into two principles: 1) the development of variation within species, and 2) the process of natural selection. Variation, also called mutation, means that some members of a certain species may develop a small genetic variation in relation to each other by chance. Natural selection, often better known as 'the survival of the fittest', means that different variations within a species may respond differently to their environment. If one particular variation seems to be better equipped for survival than another variation, then the stronger variation will develop further. In time, these variations may develop into entirely new species that are better equipped to survive.

Based on this principle, Darwin developed the theory that all species originally descended from one species and that over millions of years this one species gradually developed into the multitude of species as we know them today. Indeed, gradual development based on natural selection of variations within species is the core of Darwin's theory of evolution. One of the consequences of this theory is that the complexity of living organisms is not the result of intelligent design, but of gradual, accidental mutations and variations. It is, therefore, a concept that fits perfectly into the materialistic worldview.

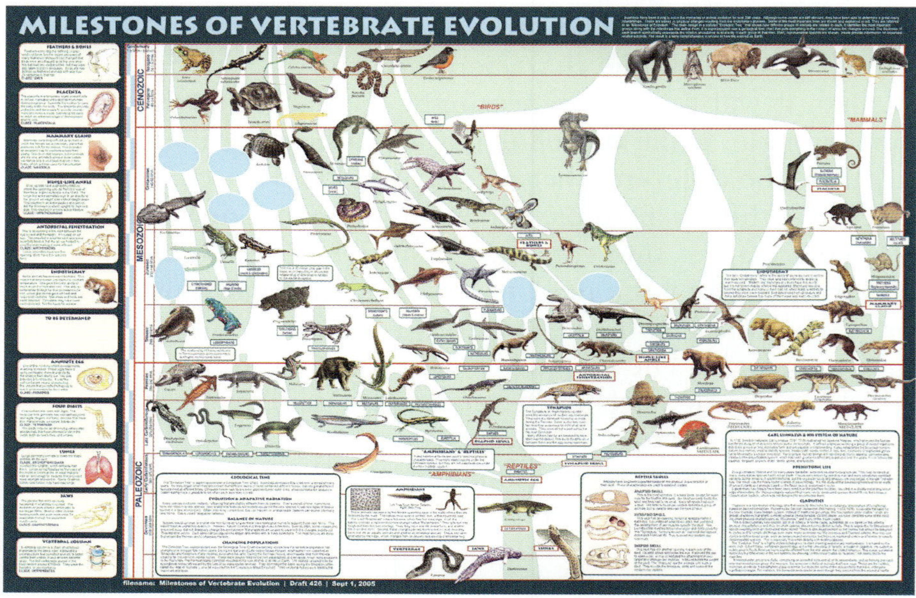

Overview of the evolution of the vertebrates. Scientists have, over the course of the years since Darwin published his theories, created vast overviews in which the evolution of species is shown schematically.

Modifications of the Theory

The theory of evolution has, since its launch in 1859, gone through some important modifications. One of the most important modifications is called Neo-Darwinism, which was a response to Darwin's erroneous view regarding heredity. While Darwin believed in the inheritance of acquired characteristics, Mendel had shown that genetic material is actually passed on via fixed laws. Furthermore, after the revolutionary discovery of DNA in 1953, science gained a notably better insight into genetics. This discovery marked the beginning of the era of molecular biology, leading to the subsequent affirmation of Mendel's doctrine whilst providing a much clearer explanation of the mechanism through which hereditary material is passed on. LaMarck's theory was immediately dismissed and Mendel's theories became integrated into Darwinism. The integration of these new insights regarding heredity led to the development of what was labelled Neo-Darwinism in the second half of the 20th century.

The second major modification was the discovery in the 1970s that the notion of the gradual development of species turned out to be incorrect. One of Darwin's heralded achievements was that he had apparently gathered empirical evidence for the gradual development of species based on his research into the fossil record. And initially, it had indeed appeared from the fossil record that there was some sort of gradual process. However, later research showed that there were enormous gaps in this theory. In 1972 two well-known palaeontologists, Stephan Gould and Niles Eldredge, published an important adaptation of the theory of evolution, called "Punctuated Equilibrium". According to their theory, there was simply not enough empirical evidence in the fossil record to support the theory that species had developed through a gradual process. Instead they proposed that evolution must have taken place in short, intense moments of great change.

Niles Eldredge, a palaeontologist of the American Museum of Natural History in New York, stated:

"The patterns of which it was claimed in the past 120 years that we should discover them, are simply not there." [34]

Stephen Jay Gould, a palaeontologist from Harvard, and co-author of the Punctuated Equilibrium stated the following:

"The fossil record with its abrupt transitions does not offer support for gradual development... All palaeontologists

know that the fossil record offers very few indications for transitional forms; transitions of one species to another are characteristically abrupt." [35]

Criticism of the theory of evolution

Nevertheless, Gould and Eldredge continued to support the theory of evolution, even though the principle of gradual development (a crucial part of Darwin's doctrine) turned out to be incorrect. They stayed loyal to Darwin's doctrine, but not all scientists are that loyal. There is a growing discomfort amongst scientists and an increasing amount of criticism of the evidence, the verifiability, and the scientific soundness of the theory of evolution. Today, there are many scientists that reject the theory of evolution as unscientific and not verifiable. The Intelligent Design movement is clear about its rejection of Darwinism and Neo-Darwinism. One of the most important arguments that they use is the teleological argument, as was extensively discussed in the previous chapter. However, other parts of the theory of evolution are also subject to serious criticism and doubt, such as the fossil record, the theory of mutations and heredity, as well as the explanations on the origin of life.

The fossil record

Coming back to the issues surrounding the fossil record. Darwin had proposed that all species originated from one common ancestor and that this ancestor had evolved by means of small, gradual, and marginal changes in its organism. The problem is that nature as we know it today provides a completely different picture. The different species of plants, animals, and humans are characterized by clear and distinct mutual differences and absolutely nowhere do we find a smooth transition showing the progression of lower life forms into higher life forms in which all in-between and intermediary links and steps are visible. Robert Carroll, a known scientist in the area of the theory of evolution, admits this in his work *Patterns and Processes of Vertebrate Evolution*:

"Although an almost incomprehensible number of species inhabit earth today, they do not form a continuous spectrum of barely distinguishable intermediates. Instead nearly all species can be recognised as belonging to a relatively limited number of clearly distinct major groups, with very few illustrating intermediate structures or ways of life." [36]

Since we cannot find any intermediary species today, it is presumed that evolution must have taken place in the past and that these intermediary life forms existed in the past. Since the enormous amount of species that currently exist is estimated at more than 8 million species, there must have literally existed millions of intermediary life forms, especially once you consider that evolution took place over extraordinarily long periods of

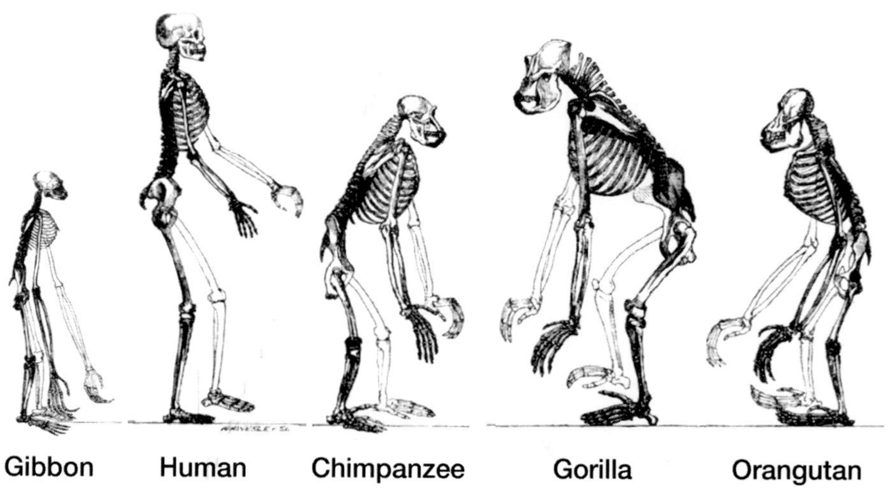

The skeleton of man next to that of other primates. It illustrates the similarities between species. On the other hand, we repeatedly see that species are strictly defined with clearly distinct characteristics.

time of tens of millions of years. The intermediary forms are described as transitionary forms by evolutionists, and based on evolutionary logic, the number of transitionary forms would have to have been far larger than the actual number of life forms that exist today. Indeed, Darwin himself accepted this as he writes in the *Origin of Species*:

"If my theory be true, numberless intermediate varieties, linking most closely all of the species of the same group together must assuredly have existed…. Consequently evidence of their former existence could be found only amongst fossil remains." [37]

Therefore, it came down to the fossil record to provide the proof which could address these concerns. And in fact, it has provided proof, just not the one Darwin had been hoping for. Almost without exception, the intermediary forms or transitionary forms are absent from the fossil record. Whilst Darwin had been painfully aware of this fact and was very worried that this would undermine his theory, he had hoped that future, more detailed research would reveal their presence. He wrote the following about this in the chapter titled 'Difficulties with the Theory':

"…Why, if species have descended from other species by fine graduations, do we not see everywhere innumerable transitional forms? Why is not all of nature in confusion, instead of the species being, as we see them, well defined?… But, as by this theory innumerable transitional forms must have existed, why we do not find them embedded in countless numbers in the crust of the earth?….. But in the intermediate region, having intermediate conditions of life, why do we not find closely linking intermediate varieties? This difficulty for a long time confounded me." [38]

Since Darwin, many palaeontologists have done a lot of research hoping to find the intermediary forms, or the 'missing links' as they are popularly called. However, they have not succeeded. Robert Carroll admits that Darwin's optimism in finding the missing links was unfounded. Contrary to the expectations of the evolutionists, when fossil remains have been found, the evidence appears to show that life on earth had appeared suddenly, completely formed. Robert Carroll explains:

> *"Despite more than a hundred years of intense collecting efforts since the time of Darwin's death, the fossil record still does not yield the picture of the infinitely numerous transitional links that he expected."* [39]

The biologist Francis Hitching, writes the following in his book *The Neck of the Giraffe: Where Darwin Went Wrong*:

> *"If we find fossils—and if Darwin's theory was correct—then we can predict what the rock should contain; finely graduated fossils leading from one group of creatures to another group of creatures at a higher level of complexity. The 'minor improvements' in successive generations should be as readily preserved as the species itself. But this is hardly ever the case. In fact, the opposite holds true, as Darwin himself complained: "innumerable transitional forms must have existed, but why do we not find them embedded in countless numbers in the crust of the earth?" Darwin felt though that the "extreme imperfection" of the fossil record was simply a matter of digging up more fossils. But as more and more fossils were dug up, it was found that almost all of them, without exception, were very close to current living animals."* [40]

In the last 40 years, there has been an increasing amount of doubt about the fossil record as empirical evidence for the theory of evolution. The former cited scientists Gould and Eldredge therefore formulated their theory of the Punctuated Equilibrium. But this theory has also been criticized, especially since the theory is not verifiable.

Mutations

Given the fact that LaMarck's doctrine regarding the inheritance of acquired characteristics was incorrect, a new theory was now needed which could support Darwin's theory of evolution. Neo-Darwinism found an apparent solution based on what geneticists call mutations. Mutations are small random alterations in the genetic code of an organism, and based on this principle, evolutionists have assumed that certain mutations could lead to evolutionary improvements of a species. In theory this solution sounds perfect, except that in reality, the opposite appears to be true. Rather than leading to evolutionary development, genetic mutations are —almost without exception—degenerative.

Biologist H.J. Muller, who won the Nobel Prize for his work regarding mutations, observed the following:

"It is entirely in line with the accidental nature of mutations that extensive tests have agreed in showing the vast majority of them detrimental to the organism in its job of surviving and reproducing, just as changes accidently introduced into any artificial mechanism are predominantly harmful to its useful operation.. Good ones are so rare that we can consider them all bad." [41]

The American pathologist David A. Demick states the following in a scientific article about mutations:

"Literally thousands of human diseases associated with genetic mutations have been catalogued in recent years, with more being described continually. A recent reference book of medical genetics listed some 4,500 different genetic diseases. Some of the inherited syndromes characterized clinically in the days before molecular genetic analysis (such as Marfan's syndrome) are now being shown to be heterogeneous; that is, associated with many different mutations......... With this array of human diseases that are caused by mutations, what of positive

effects? With thousands of examples of harmful mutations readily available, surely it should be possible to describe some positive mutations, if macro evolution is true. These would be needed not only for evolution to greater complexity, but also to offset the downward pull of the many harmful mutations. But, when it comes to identifying positive mutations, evolutionary scientists are strangely silent." [42]

The fruit fly experiments that evolutionists have performed during the last 60 years are well known. Since fruit flies reproduce very quickly—every 11 days to be precise—mutations should become quickly visible. However, despite extensive experiments over many years, new species have not come into being. Generations of fruit flies have been subject to the most barbaric circumstances, such as being exposed to extreme temperatures, radiation, or aggressive chemical substances. These miserable circumstances produced all sorts of mutations within the poor fruit flies, but none of these mutations have appeared to be in any way useful. To the contrary, they have all mostly been destructive such that the fruit flies could barely survive outside of their test tubes. Mutations, therefore, do not offer a solution in the attempt to explain the mechanism of evolution and without variation or mutation there is no evolution.

The origin of life

Another key point of criticism of Darwin's theory is related to the explanation of the origin of life itself. Darwin's theory primarily focuses on the development of existing living organisms, from unicellular creatures into multicellular organisms. While his theory implied that life originated from dead matter, Darwin did not give a detailed explanation of this. In general, he suggested that the first living cells would originate from natural chemicals and that the 'the first life form had maybe originated from a small, warm pond'. At this point, we should note that Darwin was not at all aware of the complex structure of the cell, the foundation of organic life. The existence of cells was known, but scientists at the time had no clue whatsoever of its microscopic and infinite complexity. On the contrary, cells were regarded as rather simple, gelatinous chemical compounds that could easily have developed from the correct combination of inorganic matter. The 19th century biologist and popular exponent of Darwin's theory, Ernst Haeckel, believed that a cell was

"a simple small cluster of an albuminous combination of carbon…" or a 'homogeneous drop of protoplasm." [43]

It goes without saying that this description was a colossal underestimation of the complexity of a cell. As has been stated previously, the theory of evolution implies that the origin of life must be explainable on the bases of natural, material processes. If there were no 'natural and material' explanations for the origin of life from dead chemicals, then this would have enormous consequences for the theory of Darwin. In the 1920s, the Russian biologist A.I. Oparin and the British biologist J.B.S. Haldane developed a theory regarding the origin of life based on the theory of evolution. Oparin and Haldane hypothesised that millions of years ago the chemicals that were produced in the primitive atmosphere of the earth dissolved in the primordial seas to form a sort of 'hot, watery primordial soup'. They then suggested that electrical charges generated by the atmosphere would strike this primordial soup eventually leading to the creation of the first living cells. The idea of the primordial soup appealed to the imagination of many scientists and it also become a popular concept. However, it was only an unproven hypothesis. In 1953, the scientists Stanley Miller and Harold Urey did an experiment to show that life could be created from chemicals. The scientists created a primordial soup-like substance and a replica of the primitive atmosphere on earth in a laboratory. Methane gas, ammonia, hydrogen, and water were then bombarded with electrical charges

The experiment did indeed produce certain chemical

substances, amino acids that are part of proteins, and some other organic compounds that can be found in living cells. Initially, therefore, scientists were enthusiastic about the outcome of the experiment since it had showed that some material found in living cells could originate from chemicals, and that this could even happen in an environment that resembled the primitive circumstances of a young earth. However, it became apparent from later research that the simulation of the primitive earth atmosphere that Urey and Miller had used in their experiment was incorrect. Experiments that were subsequently based on the correct atmospheric conditions completely failed and did not produce even one amino acid. Gradually, the experiment lost its

Stanley Miller

persuasive power amongst scientists and many of them rejected it all together. Furthermore, the few chemical materials that had been created in the Urey-Miller experiment still only represented a fraction of a fraction of the total molecular structure of a cell. The entire process, from Oparin to Urey-Miller, was a dead end. Nevertheless, the experiment is still to this day published in study books at schools and universities to show that life can

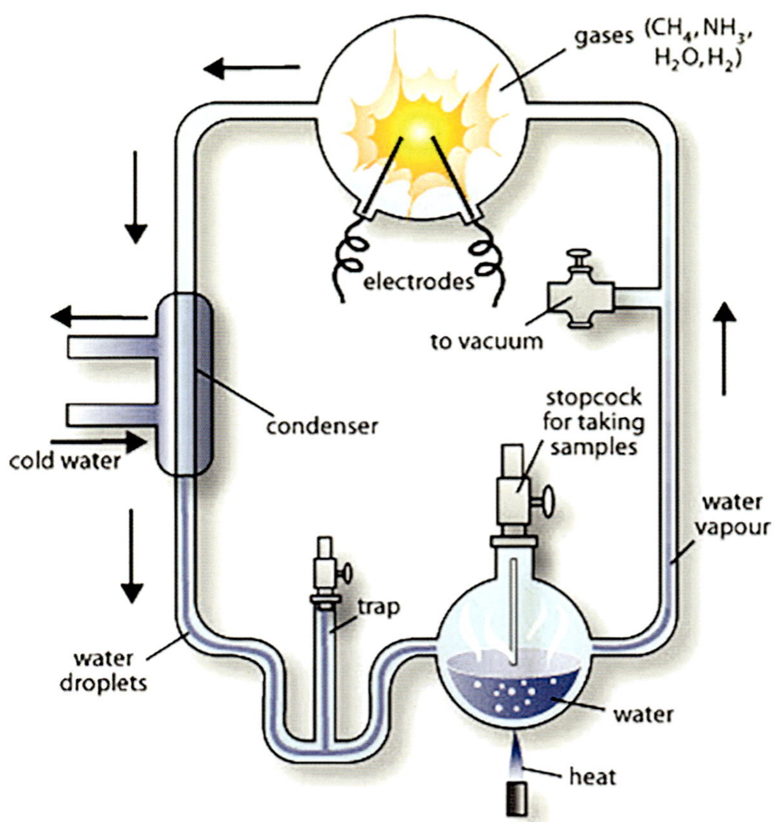

A schematic overview of the experiment that was conducted by Stanley Miller and Harold Urey in 1953. They combined a number of basic chemicals and simulated the conditions of the atmosphere of the earth as it would have been billions of years ago. Later on, it appeared that the wrong basic assumptions were used with reference to the composition of this primitive atmosphere. At any rate, the experiment did not provide convincing evidence that the first living cell could have come into being this way.

originate from chemicals, which is unproven and incorrect. The German researcher Klaus Dose wrote in 1988 that the current theory concerning the origin of life "…is a scheme of ignorance. Without fundamental new insights into evolutionary processes. [44] This ignorance will continue. The *New York Times* scientific journalist Nicholas Wade wrote in June 2000:

"Everything that has to do with the origin of life on earth is a mystery and it seems that the more we know, the more acute the puzzle becomes." [45]

Whilst the origin of life was a mystery in Darwin's time it is an even bigger mystery today. Scientists have never even come close to creating just one living cell in a laboratory by combining normal, non-living molecular compounds. Despite the fact that we know far more about living cells today, especially after the discovery of DNA and the molecular structure of the cell. Such a failure must automatically create serious doubt about the theory of evolution and the claim that life and the development of living organisms can be explained based on so called natural, meaning material and chemical processes alone.

The complexity of living organisms

In the previous chapter on the teleological argument, we extensively discussed the complexity of living organisms and how this complexity is so great that their creation and existence cannot be attributed to pure chance. It is interesting, therefore, to note some remarks that Darwin himself made regarding the issue of the development of complex organs.

"If it could be shown that a complex organ would not have been able to develop by means of numerous, successive, small adaptations then my theory would completely fall apart." [46]

The problem for Darwin is that there are indeed lots of organs that meet this qualification. A much-cited example in evolutionary literature is the human eye: a very complex organ. It is impossible to imagine an eye that is only a quarter or three-quarters complete but which functions properly. All parts of an eye have to function perfectly and they have to be in their proper place, otherwise the eye as a whole does not function. The process of natural selection, however, is a gradual process in which each organ in every phase of its development has to function properly. Each subsequent phase in the development has to be a bit better and more efficient than the previous phase. Only then does it have a better chance of surviving and consequently a better chance of developing into something new. Darwin himself made the following remark about this:

"To think that the eye could have developed by means of natural selection seems to be—and I frankly admit this—completely absurd" [47]

This issue regarding the development of complex organs is only a part of the teleological argument which leads to one compelling conclusion. Given the unimaginable complexity of living organisms, the theory of evolution is simply not supported by scientific evidence, in fact it demonstrates it to be an unworkable theory. Furthermore, as we have shown in our previous chapters, it is indefensible based on mathematical logic.

Why is evolution so popular?

Despite these problems, the theory of evolution is still the preferred theory for the majority of scientists. This is mainly because of the fact that the theory supports materialism and therefore fits perfectly in this worldview that is still dominant. Evolution implies that matter is the origin of life and of living organisms. It furthermore implies that the complexity of living organisms and signs of design are the result of pure chance and not a higher intelligence. The theory of evolution made God superfluous thereby perfectly aligning itself with materialism and atheism. The success of the theory is also due to the fact that critics of the theory of evolution often have a religious, fundamentalist vision, which makes a scientific and philosophical discussion very difficult. It is, for example, very hard to defend the position that the Earth and the universe were created by God just 6,000 years ago. This is sometimes called 'young earth' creationism, based on the literal interpretation of Genesis. Against the background of an overwhelming amount of data and proof about the age of the universe, such a point of view cannot be maintained.

Furthermore, references to religious dogmas revive old frustrations. It reminds many of an age in history where scientists, intellectuals and people who had different opinions were persecuted by the Church. In addition, many scientists are not aware of the fact that apart from the Christian young earth creationism there are other forms of Intelligent Design, and also versions of creationism that have a completely different approach. This is, as such, not a criticism of Christianity. There are, however different schools of thought within the Christian religion—including various fundamentalist movements. As we know these fundamentalist movements have, especially in the past, caused much damage and have not led to enlightenment and knowledge.

Apart from scientists, the theory is also very popular with the general public. This popularity is not only based on purely scientific arguments, but even more so on the notion that evolution is associated with progress. People are attracted to progress. The social, scientific, technological and industrial revolutions of the last 200 years have resulted in what sociologists call 'progressive thinking'. Because of these progressive social and economic developments, we can 'experience' evolution. Progress is visible and tangible and supports the thought of evolution even though this kind of progress has nothing to do with biological evolution. At its core, the theory of evolution offers a vision of humanity that is constantly developing with no end or limits. We can evolve perpetually to unknown heights with no limits in sight. This thought has provided a great deal of inspiration to the fertile imaginations of science fiction writers and offers an image of a future full of unknown and unlimited possibilities and promises.

The concept of the Übermensch is a good example of this and a clear product of the theory of evolution. Through continuous evolution, we could—in the end—gain dominion over matter and the universe. We could ultimately even become God or some type of godlike being. After all, evolution has no limits; everything is possible. Implicitly and explicitly this thought plays an important role. The Nazi ideal of the Übermensch was clearly inspired by Darwinian thoughts. Hitler did not believe in anything but himself and in Darwinism. This thought of an unknown and unlimited future with unlimited possibilities is very attractive, even if we, as an individual, may not be able to experience this development. By the time the human race has evolved into the Übermensch, we will have died and by the time that we—after millions of years of evolution—have achieved the Divine status, we will absolutely and definitely be dead. We will not personally experience it, but the human race and future individuals could experience this. This is a challenging and attractive thought.

The social influence of Darwinism

Darwin and Hitler

In this context, it is interesting to analyse the influence of Darwinism on society and to record the social and political consequences of the theory of evolution. And this influence is enormous; no other theory has influenced the world view of so many people. There are two very remarkable examples of ideologies that were inspired by Darwinism and material atheism, namely Nazism and Marxism-Leninism

Nazism is especially deeply rooted in the theory of evolution where the element of 'survival of the fittest' plays a crucial role. Marxism and its propagandists were also clearly inspired by Darwin. Darwinism gave scientific support to the materialistic, atheist vision of Marxism. Both ideologies had a devastating effect on society. The famous psychologist Eric Fromm made the following observation:

"The religion of social Darwinism belongs to the most dangerous elements of the late 19th century thought. It propagated merciless nationalistic and racist egotism and elevates this to a moral standard. If Hitler believed in something at all, then it would have been the laws of evolution that justified his actions and they sanctified in particular his cruelties." [48]

There is no doubt at all that evolutionism, the struggle for survival and the survival of the fittest were deeply rooted in the thoughts of Adolf Hitler. Hitler is regarded by many historians as the cruellest man that ever lived, however others believe that

Adolf Hitler

this dubious honour could be attributed to Joseph Stalin.

Stalin was a fanatical materialist and atheist, a follower of Lenin and Marx, and an adherent of social Darwinism. When he studied at the seminary at the age 19 to become a priest, he read Darwin's *On the Origin of Species* and became a convinced atheist. Even though Marxism does not have a direct relation to Darwinism, many of Darwin's ideas do match the Marxist worldview. Primarily Darwinism supports the materialistic and atheist basic assumptions of Marxism. Marxists believed that the propositions of Darwinism provided a scientific foundation for their atheistic and materialistic world view. The Marxist class struggle, accompanied by violence and revolution, is in many ways comparable to the process of natural selection and the struggle for survival, but then applied to social classes. Stalin's fellow revolutionary fighter Leon Trotsky (1879 -1940), initially the designated successor of Lenin, was even more outspoken in his support for Darwin. In 1901, when Trotsky was imprisoned in

Joseph Stalin

Odessa because of his revolutionary activities, he started reading *On the Origin of Species*. Years later he wrote:

"Darwin destroyed the last of my ideological prejudices. While I was in prison I suddenly felt a solid and scientific surety... Darwin stood right in front of me as a powerful gatekeeper at the entrance of the temple of the universe..." [49]

Decades later, in 1940, Trotsky was killed by Stalin, since Stalin regarded him to be a dangerous competitor. For Stalin, murdering his opponents and fellow fighters who he believed to be too powerful, and therefore dangerous, was a matter of inevitable necessity. A necessity comparable to what happens in nature; natural selection and the hard struggle for survival. There is no doubt that the influence of Darwinism on Marxism in general—and on Lenin, Stalin, and Trotsky especially—was enormous.

While Hitler was not necessarily an atheist as Stalin was, he

Leon Trotsky

was a fanatic adherent of social Darwinism and of the racist ideas which were based on evolution.

Hitler wrote in *Mein Kampf*:

"He who wants to live has to fight; he who does not want to fight in this world where permanent struggle is a law of life, does not have the right to exist." [50]

The Nazis believed that by extinguishing the inferior, weaker races, they were only helping nature by speeding up the inevitable process of natural selection. This assistance to nature was presented as a religious duty and sacrifice, a major offering to nature and humanity. The Nazis had succeeded in constructing a racial doctrine with all sorts of gradations of superior to inferior, based on Darwinian principles. Eugenics, the science dealing with ennobling races, was a cornerstone of Nazi ideology. The term "eugenics" was invented by a cousin of Darwin in 1865, six years after *On the Origin of Species* was published. The Nazis used the tenets of eugenics to murder tens of thousands of physically and mentally handicapped people. This does not imply, though, that Darwin or his cousin are responsible for the atrocities of the Nazis, or that they would have agreed to such a course of action. Yet, it is an inevitable result of Darwinian thought, if this thinking is applied and interpreted strictly and stringently. The Nazis just applied the process of natural selection without compromise, unscrupulously, and mercilessly. In a Nazi propaganda film from 1937 titled 'Victims of the Past', the audience is shown a mentally handicapped person and the voice-over explains:

"During the last decades, humanity has sinned terribly against the laws of natural selection. We have not only kept life alive that is unworthy of living, but we have even given it the opportunity to multiply itself." [51]

In his book *The Descent of Man*, Darwin himself articulated the practical effect of natural selection and he predicted—unintentionally—the following:

"In the future and measured not very far away in the future, civilized human races will undoubtedly replace and eliminate the barbaric and uncivilized races."

Again, Darwin would not have said this with extermination camps in mind, and he would also not have thought that people would actually start implementing and executing the principles of natural selection themselves. However, this is the consequence of a rigorous and fundamentalist interpretation of the theory of evolution, natural selection, and the law of power and the law of the strongest. The scientific arguments based on the theory of evolution were very useful to convince intellectuals of the truth of Nazism. It made the ideology more credible and it had significant PR value. This, however, did not change Hitler's negative view of intellectuals and scientists. The Nazi ideology was an incoherent ragbag of different movements, such as 19th-century Romanticism, various mystical and esoteric theories mixed with Nordic mythology, and, of course, the theory of evolution. The Nazi ideology was at its core anti-intellectual and anti-scientific. It is interesting, in this context, to consider the following quotes of Hitler:

> *"We are now at the end of the Age of Reason," Hitler stated towards Hermann Rauschning, a former friend of his. "The intellect has grown autocratic and it has become a disease of life. We must mistrust the intelligence and our conscience, but we should have confidence in our instincts." He further said: "Trust your instincts or your feelings or whatever you want to call them."* [52]

Hitler also made the following statement in relation to the upbringing of children and education:

> *"I want a violent, dominant, intrepid, and brutal youth… I do not want intellectual training. Knowledge will only ruin my young men."* [53]

All these statements fit the image of a world driven by evolutionary instincts to survive, where 'might makes right' and the power of the strongest rules which principles became the moral standard. Since Nazi ideology was irrational in its core, it did not have any interest in presenting a logical and consistent worldview. The Nazi ideologists used science the way it suited them best, and if it supported their practical and political goals. If the image of a society driven by instincts could be founded on a scientific theory with an intellectual veneer, then that would be politically useful. Hitler had to build support in society, especially in the beginning phase of the rise of Nazism when the support of industrialists, intellectuals and other influential people was absolutely necessary. Social Darwinism gave him the means to do so. Apart from that,

Hitler was indeed convinced of Darwinism as Fromm stated; this was about the only thing that he really believed in.

Darwin's theory of evolution was used by the Nazis as scientific justification for their extreme points of view and their criminal deeds. Natural selection became the foundation for a racist ideology that became the core of the Nazi ideology. According to this racist ideology, it was necessary to eliminate the weaker races and thereby free the genetic pool from negative influences. For this reason, a religion like Christianity was considered to be an anti-religion and destructive for human development. Christianity preached mercy and protection of the weak and the helpless. The Nazis on the other hand, believed that protecting and helping the weak would increase the amount of weak people in the world. As a consequence the genetic pool would be contaminated with an increase of weaker genes. Soon after the assumption of power by the Nazis in 1933, their ideology was put into practice by systematically murdering physically and mentally handicapped people, including children. According to Nazi philosophy this was no different than applying the laws of evolution and natural selection. Later on, these practices were expanded to the systematic extinction in extermination camps of, according to the Nazis, inferior races and communities. This basically meant everyone who was not German with blond hair and blue eyes. The Jews and the Slavic people of Eastern Europe and Russia in particular were victims, but also gypsies, homosexuals, and other minorities who did not fit the picture of the Aryan Übermensch. The end purpose of this 'vast purification' and this horrific killing was the establishment of the 1000-year Great German Empire, the Promised Land. This empire would stretch from the coasts of the North Sea up to the Ural Mountains, inhabited by the Aryan Übermensch with the Führer as its new God. Total world domination would be within reach and would only be a matter of time. Evolutionary forces would reach new heights and humanity would start a new era. It would be an era dominated by the Aryan Übermensch as the pinnacle of evolution; that is what they hoped for, but as we know, fortunately, history decided differently at this point.

Evolution is only a theory

The theory of evolution is only a theory for which the scientific solidity is extremely dubious. The arguments and evidence against the theory vastly outnumber the evidence in favour of it. Real, solid proof in favour of the theory cannot be found. In nature, we do not observe examples of evolution, meaning examples of a species developing into a new species. We also do not observe the development of intermediary forms. Nor has there ever been a successful experiment in a laboratory that has created a new species or even just a single living cell.

Various other inconsistencies with the theory of evolution have been discussed. According to the modern laws of genetics, we see that, in reality, species are rigid and inflexible and that they stick to clearly defined characteristics, contrary to what should be expected according to evolution. Furthermore, the fossil record, once one of the strongest aspects of Darwin's theory, also appears to be incomplete and inconsistent and has actually become an argument against his theory rather than one in favour of it.

The Darwinian explanation for the origin of life is also a major problem, quite simply because there is no scientific explanation. It remains a giant enigma, even more so today than in Darwin's day given our insights into the true complexity of just one single cell. The basic proposition that life has appeared from dead matter runs contrary to everything that we witness, observe and can scientifically verify. In all of recorded human and scientific history no one has ever experienced or observed life appearing from dead matter, even using all of our technology and accumulated knowledge to artificially speed up such a process. Yet, every day we witness that life comes from life; in fact, life is only ever passed on from another already living source. This is a verified and conclusive fact yet atheism and materialism argue the opposite; that life must have somehow appeared from dead matter without any evidence for this whatsoever.

Yet, the biggest problem for the theory of evolution is the statistical impossibility that it could ever have taken place. The complexity of living organisms, whether we are talking about the development of just one cell or the subsequent development from one cell into a multicellular organism, is so unimaginably great that random chance cannot possibly have played a part in this. The teleological argument is unambiguous and conclusive from any rational and scientific point of view. The probability that life came into being out of matter by pure chance is zero, and the probability that one single cell could randomly evolve into a human being with 80 trillion cells is also zero.

What we can determine is that Darwin's theory is based on assumptions with a very limited amount of evidence and proof. The success of the theory is mainly due to the necessity for such a theory. Without evolution, one would have to revert to creationism and theism, for there are, after all, no other options: living organisms have either come into being by pure chance or by intelligent design. However, the implications of such a choice go too far for many scientists as the theologian Dr. Clark Pinnock observed:

"The reason that evolution is considered to be true and is taught as the truth is not because of the evidence for it, but because of its necessity." [54]

Since Dr. Pinnock is a theologian, he could be accused of partiality. However, one can hardly suspect the formerly cited Fred Hoyle of being partial, since he was an atheist and he believed in evolution until new scientific insights made him change his mind. This led to the formerly cited and remarkable quote:

"Personally, I have little doubt that scientific historians of the future will find it mysterious that a theory which could be seen to be unworkable, came to be so widely believed." [55]

In the following chapter, we will discuss the scientific developments that contradict materialism one by one and in more detail. Accordingly, we will also examine what the consequences are and how a proper and correct interpretation can be given to a scientific, not materialistic, worldview.

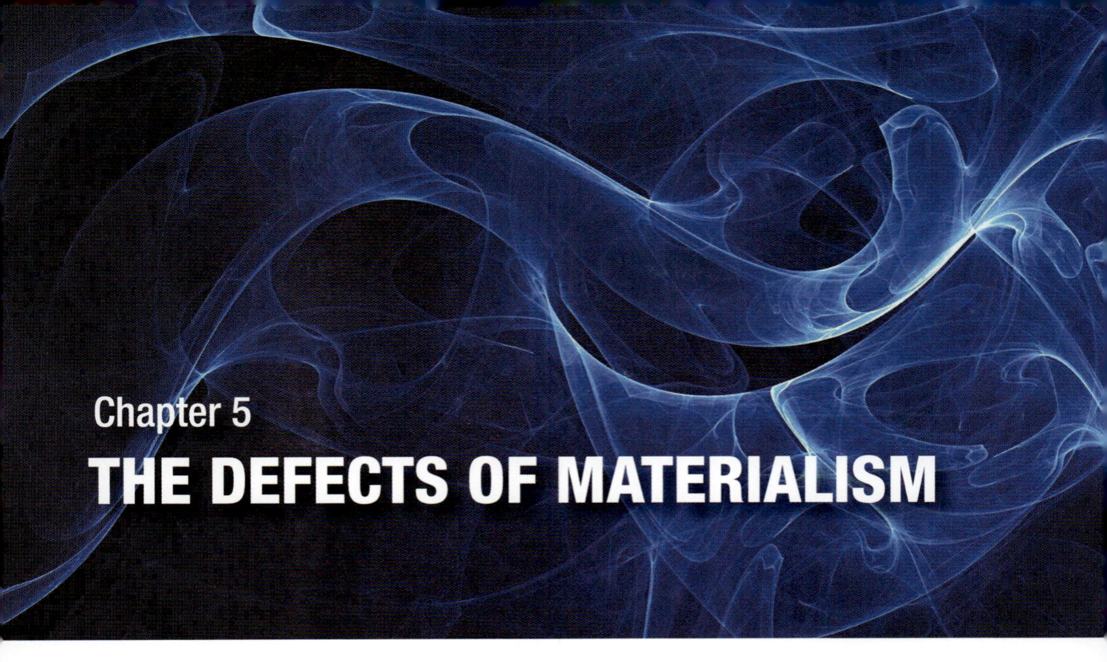

THE DEFECTS OF MATERIALISM

"Everyone who is seriously involved in the pursuit of science becomes convinced that a spirit is manifest in the laws of the universe —a spirit vastly superior to that of man"

Albert Einstein, 1936 [56]

"A little bit of science leads to atheism, but large quantities of it bring us back to God"

Francis Bacon – British philosopher and scientist
(1561-1626)

More defects?

In Chapter 2, 'the Teleological Argument', it was demonstrated how important parts of materialism as worldview have become inadequate and unworkable. In this chapter, additional evidence will be offered that demonstrates the inadequacies of materialism, specifically the notion that matter is the only existing reality in the universe. Additional evidence will also be offered that materialism, with its obedient servants chaos and chance, cannot possibly explain the level of organised complexity that we observe in the universe. This holds for both the biological complexity of living organisms as well as the cosmic complexity of a fine-tuned universe.

In the previous chapter a short historical overview was provided of the origin of science and the resulting development of materialism as a worldview. In particular, we discussed the logical progression of scientific discoveries and their related philosophical insights. From this we could clearly identify a logical pattern that led to the development of materialism as a worldview up to the beginning of the 20th century. The historical overview shows however, that from the beginning of the 20th century seismic changes are starting to occur. New scientific

On the previous page a part of a well-oiled machine is depicted. Materialism regards the universe as a machine in which parts function due to direct physical contact with other parts. This physical contact is the mechanism through which movements, change and forces are transferred onto each other. The Newtonian image of the universe is founded on these 'mechanical' principles. The New Physics, however, has revealed a universe in which such mechanical principles are totally inadequate to explain the functioning of the universe. Force fields — or energy fields — operate in a mysterious, non-mechanical way, in which objects influence each other at a distance and without 'material' intervention. Furthermore, the interactions and structures of energy fields, atoms, and sub-atomic particles are so complex that the existence of a coordinating intelligence is unavoidable.

discoveries and insights such as relativity, quantum theory, and Big Bang cosmology, contradict the materialistic world view each in their own way. These new insights extend to five areas in total of which the first area—the explanation for cosmic fine-tuning and the resultant organized complexity in the universe—has been extensively discussed. This chapter will focus on discussing the other four areas in detail. In summary these four areas are:

1. The limitations of our senses and of direct observation.

The core of this argument is that the scientific method and materialism can no longer base themselves on direct observation. Observation has become too complex and too abstract. This conclusion does not directly undermine materialism, but it does weaken the method that lays the foundation for materialism. Two important areas of research that demonstrate this argument include:

- The world of the atom
- Quantum theory

The key conclusion is that observation can no longer be detached from consideration and philosophy. As the famous astronomer Edwin Hubble stated:

"Observation always coincides with theory".

Consequently, modern physical theories have acquired a metaphysical character out of sheer necessity, where they compete with other metaphysical worldviews. Furthermore, the degree of truth is no longer determined just by experiment and observation. Observation remains fundamental, but increasingly, the degree of truth of a theory is determined by the force of logic and the internal consistency of that logic. A logic that should also be compatible with the existential and deepest inquiries that concern the universe.

2. The discovery of the existence of force fields and energy fields that do not comply with the traditional definition of matter.

This discovery leads to the realisation that the 'ordinary' natural world is extraordinary. The theory of relativity is a good example of a theory that shows the extraordinary nature of reality. The role of energy as the foundation of existence similarly leads to just such an understanding. It is primarily these three areas of scientific research that confirm the extraordinary and non-material nature of reality:

- The discovery of force fields and the extraordinary properties of energy, such as gravity and electro-magnetism.
- The theory of relativity
- The discovery that energy forms the foundation and basis of reality

3. The discovery of entirely non-observable realities, such as dark matter and dark energy

This argument is also part of the first argument, which is the limitation of our sensory perception. In the case of dark energy and matter, this limitation is even more extreme, since we cannot even speak of a partial or limited perception. Dark energy and matter have completely concealed themselves, at least up to now, from any form of perception. Their existence can only be concluded based on derivatives and assumptions.

4. The implication of the big bang theory that the material universe originated from something that is fundamentally different and non-material i.e. the singularity

Big Bang cosmology implies that at its core, matter has a non-material cause. It furthermore implies the fundamental unity of the universe, since it originated from something that is fundamentally one. Cosmology has undergone a dramatic change with the discovery of the Big Bang, and has led to an entirely different understanding of the universe. We now understand that the universe must have a beginning and an end in time, and that both the beginning and the end, each in their own way, have mysterious and supernatural implications.

All of the above-mentioned points strongly undermine the materialistic worldview, and defy the notion that nothing exists outside of matter, one of the four cornerstones of atheism. The introduction of the theory of relativity of Albert Einstein in 1905 and 1916 and the formulation of quantum physics in the 1920s have been truly revolutionary. And so was the discovery of the subatomic world by Ernst Rutherford towards the end of the 19th century. Relativity and quantum physics have been called the 'New Physics' to distinguish them from Newton's classical physics, which does not imply that classical physics is incorrect, but rather that it has turned out to be incomplete. This incompleteness refers to the relation between space and time, the relation between energy and mass, the structure of the atom and many of our 'common sense' insights regarding these phenomena. The mechanistic foundation of Newton's classical physics only provides a partial explanation of the manner in which nature and the universe present themselves to us. Finally the Big Bang cosmology implies that our universe has a beginning and originates from something that is not material, which is further evidence that materialism is limited and fundamentally incorrect.

The limitations of direct perception

Despite the success of the scientific method, scientists soon became aware of the limitations of our sensory perception. Therefore, as early as the 16th century, scientists and ingenious inventors experimented with all sorts of new devices that could push these boundaries. For example, the invention of the microscope by the end of the 16th century enabled the Dutch scientist Anthony van Leeuwenhoek (1632-1723) to discover the world of bacteria and other microscopic life. This was a revelation for scientists, which provided them with access to an entirely different world. In a similar way, and at about the same time, the invention of the telescope also led to a revolution in cosmology.

The Dutch inventor Hans Lippershey discovered in 1608 that objects could be enlarged significantly with the use of lenses. This invention was applied by Galileo in 1609, who saw the use of this new instrument as an opportunity to prove the theories of the Polish astronomer Copernicus. Copernicus had suggested, based on mathematical calculations, that the sun, rather than the earth, was the centre of our solar system. Galileo was now able to prove this theory by empirical observation with the assistance of the telescope. These inventions were the beginning of an enormous expansion of the arsenal of instruments and devices that would enable scientists to dramatically expand the process of observation. Better equipment led to new scientific discoveries and new technologies that in turn led to new and better equipment.

The biggest technological developments, however, have taken place in the 20th and 21st centuries. Electron-microscopes, radio telescopes, super computers, and particle accelerators have

Hans Lippershey

pushed the boundaries of our sensory perception to previously unimaginable limits. Gradually, the realm of the unobservable, difficult-to-observe, and indirectly observable worlds got larger and the world of the observable became smaller and smaller. Visible light, for example, represents only 0,0035% of the entire spectre of electromagnetic radiation. On the other hand, infrared or ultraviolet radiation is something we cannot see with the naked eye, but we know it exists because we have developed devices and technology such as infrared viewers that allow us to detect its existence.

In another example, using very powerful telescopes, such as radio telescopes and the Hubble telescope that is stationed in space, we have been able to see much further into space. The Hubble telescope is particularly powerful, since it is not hindered by the Earth's atmosphere. As a consequence, the images that Hubble produces are extraordinarily sharp and Hubble is therefore able to take pictures of deep space that were not accessible before. From these observations, we have discovered that there are in fact billions of other star systems

First telescope, designed by Hans Lippershey

which exist outside our Milky Way, and consequently, the size of our visible universe has become billions of light-years larger. At the same time, this implies that the ratio between what we have been able to observe in comparison to what is out there, has shifted considerably. In other words, the universe is much larger, massively much larger than we thought just 100 years ago. And as the universe continues to expand, we are becoming inversely and proportionally smaller.

The world of the atom

Whilst technology and science have greatly expanded our understanding of the universe, our observations have also become more indirect and more abstract. This is especially true of the tiny universe that we call the atom. Newton had believed that atoms were indivisible, the undividable 'rock-hard' materials of matter. The existence of these rock-hard materials matched perfectly with the mechanistic worldview of colliding objects. However, it has since become apparent that this aspect of

Karl G. Jansky Very Large Array (VLA) in New Mexico, U.S.A. The original telescope as it was invented by Lippershey made use of visible light as its information source. Visible light is only a small part of the electromagnetic spectrum that largely consists of radio waves. A radio telescope is able to catch radio waves with a huge range of wave frequencies apart from the frequencies of visible light. Therefore, the ability to detect signals from the universe is strongly enhanced. The Karl G. Jansky Very Large Array (VLA) is a radio astronomical observatory situated in Socorro County, New Mexico, in the U.S. The observatory consists of 27 independent antennas, of which each antenna has a cross-section of 25 metres and weight 209 tonnes. These are positioned as three legs of a Y (all together 21 kilometres long). By using rails, the antennas can be placed in different spots. The antennas are linked to each other and the information that each separate dish antenna catches is used to make interference patterns. These patterns give insight into the structure of sources that transmit radio radiation in the air. These are in turn mapped out by using a mathematical technique, called the Fourier transformation. The smallest resolution that can be achieved is one of 0.05 arc seconds with a wavelength of 7 millimetres.

classical Newtonian physics was incorrect, or at least incomplete. Even though we experience matter as a 'hard' and 'solid mass', the materials that are responsible for this hardness appear to be made of something totally different. This realization came in the 1890s, when radiation and electrons were discovered. These discoveries led to the unavoidable conclusion that atoms are

The Hubble space telescope circles at a distance of 569 kilometres above Earth

not indivisible and unchangeable as Newton had assumed. The atom appeared to be divisible, and it also appeared to possess a complex internal structure.

Scientist Ernest Rutherford (1871-1937), who is considered to be the father of nuclear physics, further expanded the atomic theory when he discovered the internal structure of the atom. In 1911 following numerous experiments, Rutherford reached the conclusion that the internal structure of the atom consisted of a small, positively charged nucleus, encircled by negatively charged electrons. In the 1930s, a neutrally charged particle, the neutron, was added to the nucleus leading to the now-classical atomic model: a miniature solar system, consisting of a positively charged nucleus, a proton and a neutron, orbited by negatively charged electrons.

Atoms are invisible to the naked eye, even for the most powerful of microscopes. Recently, an electron-microscope was built that is being used at the University of Victoria in Canada which is able to detect the contours of an atom.

This microscope, the Scanning Transmission Electron Holography Microscope, or STEHM, is able to enlarge objects up to 20 million times. The STEHM weighs about 7 tonnes, is 4.5 metres high, and is currently the most powerful microscope on earth. In comparison, the most powerful microscope that uses light can only enlarge objects up to 2000 times. The reason that electron-microscopes are much more powerful than optic microscopes is that an electron is 100,000 times smaller than a light photon. The electron-microscope was invented in 1931 by a

An atom nucleus with electrons circling around it.

The Large Hydron Collider (LHC) in CERN. The illustration above shows a part of the 25-kilometer-long tube in which sub-atomic particles are accelerated up to the speed of light. In doing so they collide with other sub-atomic particles occasionally causing new particles to manifest themselves for perhaps just a millionth of a split second.

German scholar, Ernst Ruska, and has been improved enormously since then. It has given scientists access to the extremely small micro-nature of our world and the unimaginable complexity hidden within it.

Subatomic particles, however, remain outside the range of even electron microscopes. In order to construct a realistic image of the atomic and subatomic world, scientists had to resort to the use of exotic and truly gigantic machines, such as particle accelerators. At CERN in Switzerland, scientists have built the biggest particle accelerator in the world. This accelerator has a circular length of about 25 kilometres, in which particles are accelerated to the speed of light. At that speed, they collide with other particles and are split into even smaller, possibly particles that were previously unknown.

Nevertheless, even with such incredible technology, direct observation of the atomic universe remains well out of reach. The observations in a particle accelerator, for example, are so abstract,

Fritjof Capra

and dependent on interpretation that one can hardly speak of observation. In effect, the existence of subatomic particles cannot be determined through observation but only through indirect inferences. The following descriptions illustrate the immeasurably small size of the atomic and subatomic universe.

The diameter of an atom is about one-hundredth of a millionth of a centimetre, a miniscule size that we cannot possibly fathom. However, it is even stranger that the atom itself consists mainly of empty space, with only a very small nucleus at its centre. The nucleus of an atom is so small that it would only become visible if one were to increase an atom to the size of the largest dome of the world, namely the Saint Peter basilica in Rome, which has a diameter of 42 metres and a height of 139 metres. The nucleus of such an enormous atom would then be the size of a grain of salt. Capra describes this in his book *The Tao of Physics*:

> *"A grain of salt in the middle of the dome of the Saint Peter basilica, with specs of dust whirling around it in the vast space of the dome – this is how we can picture the nucleus and electrons of an atom."* [57]

Keeping in mind that the atom itself is only one-hundredth of a millionth of a centimetre wide in diameter, its nucleus is again one-hundred-thousand times smaller than the entire atom!

The subatomic structure is often compared to our solar system, where the nucleus is the sun and the electrons are pictured as the planets that circle around it. Up to this point the comparison is applicable, but from there on the comparison breaks down. The earth orbits the sun in 365 days, at a speed of 107,000 kilometres per hour, while an electron moves around at a speed of 965 kilometres per second, or, in other words, at a speed of 3.4 million kilometre per hour. Moreover, an electron orbits the nucleus more than 3,000 trillion times per second! These speeds all take place in dimensions that we cannot possibly fathom. And even more amazing are the speeds occurring within the tiny nucleus of the atom. According to Capra, protons and neutrons move at an even more staggering speed of 64,000 kilometres per second in a space that is 100,000 times smaller than the atom. Capra further explains that the mass of the atomic nucleus has a remarkably high density:

> *"Indeed if the whole human body were compressed to nuclear density it would not take up more space than a pin-head."* [58]

What is crucial is that the process of observation is nothing like the traditional way of observation that the founders of empiricism and the scientific method had in mind. Capra explains this as follows:

> *"The   delicate   and   complicated   instruments   of   modern*

experimental physics penetrate deep into the sub-microscopic world, into realms of nature that are far removed from our macroscopic environment, and make this world accessible to our senses. However, they can do so only through a chain of processes ending, for example, in the audible click of a Geiger counter or in a dark spot on a photographic plate. What we see or hear are never the investigated phenomena themselves, but always their consequences. The atomic and sub-atomic world itself lies beyond our sensory perception." [59]

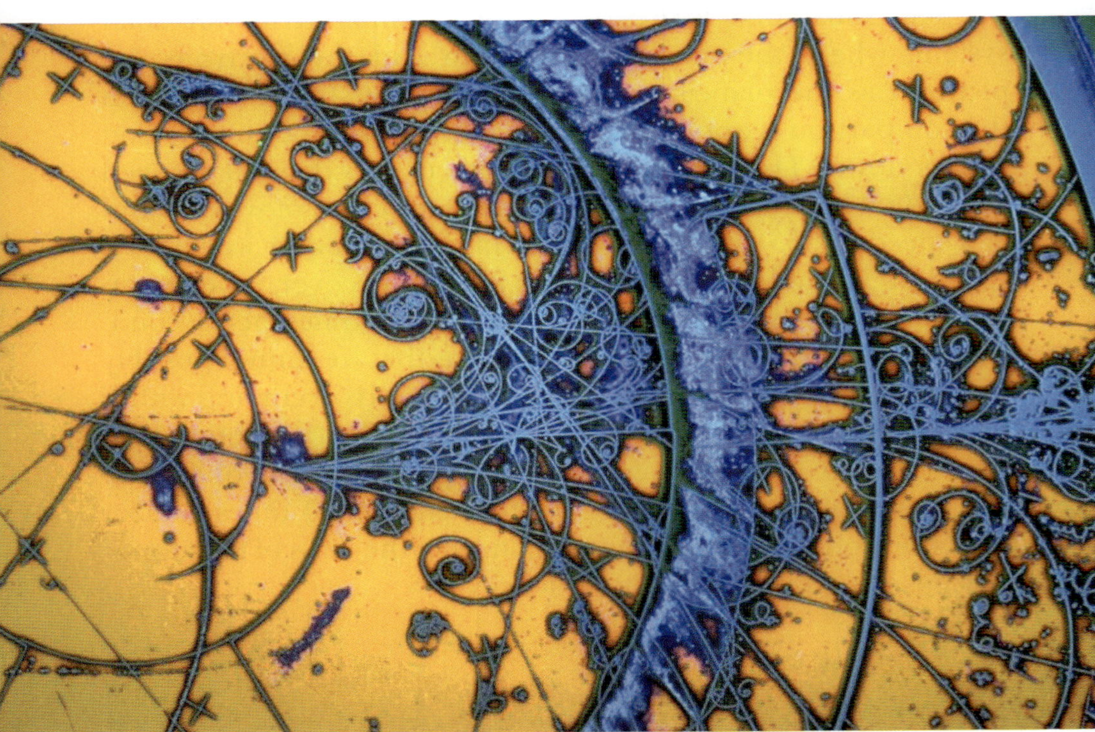

An artistically embellished illustration of the kind of observations that occur in the Big European Bubble Chamber (BEBC) in CERN. Detectors are used to measure the properties of particles; some detectors measure traces that are left behind by particles and others measure energy. CERN was established in 1954 and it is the biggest and leading research laboratory in the area of particles physics in the world. A Bubble Chamber is a container filled with extremely hot liquid hydrogen. Extreme heating occurs by a temporary, sudden decrease in pressure in the container, which causes the temperature to rise above boiling point. Each charged particle that goes through the liquid at that moment leaves a trace behind of small 'bubbles' since the liquid boils its wake. These 'bubbles' are detected as small traces, from which the characteristic path and the properties of the particle can be derived.

And he continues in the next paragraph:

"It is, then, with the help of modern instrumentation that we are able to 'observe' the properties of atoms and their constituents in an indirect way, and thus to 'experience' the subatomic world to some extent. This experience, however, is not an ordinary one, comparable to that of our daily environment. The knowledge about matter at this level is no longer derived from direct sensory experience, and therefore our ordinary language, which takes its images from the world of the senses, is no longer adequate to describe the observed phenomena. [60]

In this statement, Capra clearly formulates the dilemma of the scientific method regarding the non-perceptibility of the atomic and subatomic world.

Quantum theory

In order to describe the subatomic world and try to understand it, a number of scientists in the nineteen-twenties developed an entirely new theory: quantum theory. Quantum theory became a milestone in the development of modern physics. The founder of quantum theory is the renowned German physicist Max Planck (1858-1947) who received the Nobel Prize for his work in this area.

Other important scholars who made major contributions to the development of quantum theory were Danish theoretical physicist Niels Bohr (1885-1962), German physicist Werner van Heisenberg (1901-1976), and Austrian physicist Erwin Schrödinger (1887-1961). Based on research and observations, these scientists discovered that the sub-atomic world behaves differently from the world of matter that we know. They discovered that sub-atomic particles are not fixed objects, but abstract units, more comparable to states of energy.

The essence of quantum theory and quantum field theory is that material particles are basically manifestations of energy fields and that a sub-atomic particle is a packet of energy. This implies that sub-atomic particles sometimes behave like a particle and sometimes like an energy wave. This fact also makes it quite difficult to locate sub-atomic particles. A particle is at a location, a wave is everywhere. Werner von Heisenberg formulated in 1927 the currently famous 'uncertainty principle' where he stated, in short, that it is not possible to determine the speed and the position of a sub-atomic particle simultaneously. The reason for this is that particles at a sub-atomic level sometimes behave like waves and sometimes like particles.

Another consequence of the uncertainty principle is that the act of observation itself changes the observed object. This was first regarded as a technical problem that resulted purely from the fact that the observed phenomena were so small that observation and observational tools would disturb the observed system. In order to observe a sub-atomic system, it may be necessary to use other sub-atomic particles to make it 'visible'. These particles, however,

Max Planck

will have to disturb the system that they are trying to reveal.

Heisenberg did not state that consciousness itself could directly influence the results of sub-atomic processes, however other physicists have suggested it did after new discoveries were made in the field of quantum physics. Psychologist Jack Sarfatti wrote about this in his article "Implications of Meta-Physics for Psycho-energetic Systems":

"A very important idea regarding the development of psycho-energetic systems is that the structure of matter may not be separate from consciousness!"

An important consequence of quantum theory is that reality, at a sub-atomic level, behaves paradoxically. Particles that manifest as an energy wave, and an energy wave that subsequently manifests itself as a particle. Quantum physicists combined

Werner von Heisenberg

the terms particle and wave and coined the term 'wavicles' in order to express this paradox. Quantum theory implies a very mysterious world that is not normal but extraordinary in every way. To quote Heisenberg again:

> *"I remember discussion with Bohr which went through many hours till very late at night and ended almost in despair; and when at the end of the discussion I went alone for a walk in the neighbouring park I repeated to myself again and again the question: Can nature possibly be so absurd as it seemed to us in these atomic experiments?"* [61]

We have just alluded to the possible role that a conscious observer could play in relation to objective reality. In quantum physics this is called the Copenhagen interpretation. According to this interpretation, objective, sub-atomic reality consists of fields of energy in which all possible potential objects with their potential properties are situated. At the moment of observation, this field of potential objects and properties changes into a number of concrete and actual objects and properties that accordingly determine the nature of the observation. In quantum jargon this process of observation is called the 'collapse of the wave function'. Referring to the statement of Jack Sarfatti, the Copenhagen interpretation can lead to the philosophical view that the individual conscious observer is the creator of that which he observes. There are several philosophical schools of thought that use this interpretation of quantum physics. In later chapters, we will discuss this matter in more detail. One of the consequences of this interpretation is that it leads to bizarre paradoxes. The most well-known of these paradoxes is the paradox of Schrödinger's cat. In short, this paradox means that a cat, based on the quantum interpretation of reality, can be dead and alive simultaneously.

Above, is an image of the famous cat paradox with an explanation based on a translation of the original text of Erwin Schrödinger.

"A cat is locked up in a steel room, together with the following diabolical machine (that one has to protect from direct interference by the cat): in a tube there is a miniscule amount of a radioactive element, so little that during one hour it is possible that one of the atoms decays, but just as likely that it will not. If an atom decays, then a Geiger counter detects this and will drop a hammer, via a relay that breaks a bottle with hydrocyanic acid. If one has left this system to itself for one hour, then one can say that the cat is still alive, if in the meantime an atom hasn't decayed. The first atom that decays would have poisoned the cat. The functional condition of the entire system would express it in such a way that the living cat and the dead cat would occur mixed and simultaneously. The characteristic of such cases is that an original indeterminate state up to atomic level is translated into a gross sensory indeterminateness that can be finally determined and decided

Niels Bohr

by direct observation. If it is true that a particle does not necessarily exist until it is being observed, then it is not sure whether the hammer will ever fall—maybe it has to fall and maybe it can fall and maybe it cannot fall. Until the box is opened, we cannot tell what has happened. The extension of possibilities over the field of probability means, therefore, that as long as the box is closed, the cat can either be dead or alive. As long as no observation is possible, there is no other way to tell." [62]

Schrödinger's cat is a somewhat strange way to explain a physics proposition. First of all it is not the case that cats have a special relationship with quantum theory; Schrodinger could also have used a rat or a chimpanzee in his analogy. Schrödinger used this analogy to clarify the absurdness of the Copenhagen interpretation of quantum theory. He did not completely

Image of the famous cat paradox of Erwin Schrödinger

succeed in this, for the cat paradox started a new life of its own. Cats are known to have several lives, so perhaps Schrödinger indeed should have better used a rat in his example. His cat paradox has in any case led to many controversies in science regarding the quantum doctrine. It seems that Stephen Hawking once said: "When someone mentions Schrodinger's cat, I am inclined to draw my pistol". Even the most rational scientist can apparently loose his cool. Contrary to Schrödinger's intention, the cat paradox only made the Copenhagen interpretation more popular.

It is interesting to note at this point, that the simultaneous existence of potential substance and actual substance, as well as potential properties and actual properties within a field of energy completely matches the vision of the Indian Vedanta philosophers. This will be discussed extensively in chapter 10 where we will also refer to a paradox. This paradox relates to the simultaneous existence of actual and potential properties, including properties that are each other's opposite and therefore basically mutually exclusive and unable to coexist with each other. In Vedanta philosophy, the process of the transformation from a potential to an actual object and from a potential to an actual property is indeed generated by consciousness, but not by the individual consciousness of a specific, individual observer. This transformation is generated by the infinite consciousness and the all-encompassing intelligence of God. Within that context, Schrödinger's cat paradox indeed shows the absurdity of a worldview where coincidental individual observations create objective reality. This situation becomes even more absurd when we assume that there are different observers who make their observations simultaneously. The world would, in that case, degenerate into complete chaos of objects jumping in and out of reality without structure or coherence. It would also suggest a world in which individual observers could change the appearance of objects and their properties in accordance with their personal wishes by looking at things differently.

This vision is a form of idealism, a philosophical school that assumes that reality is a product of consciousness. The word

Erwin Schrödinger

idealism in this context should not be confused with trying to achieve an ideal. Idealism is a term derived from the word idea: thought or consciousness. Idealism is divided into two categories, subjective idealism and objective idealism. The difference is exactly what is discussed here. Subjective idealism states the proposition that the objective reality is generated by the individual subjective consciousness of one person. This will lead to the situation as described above: cat paradoxes and a chaotic reality. Objective idealism also assumes that consciousness is the creative force, but places this creative force outside the individual consciousness of a single observer. Objective idealism assumes that there is a cosmic, conscious force that creates the world as it is and that this world is a transformation of consciousness and is led by consciousness. This avoids the cat paradox and eliminates the individual observer as creator of the universe. Objective idealism is effectively one of the cornerstones of the Vedanta school as will be discussed later on in chapter 7.

The conclusion is that our ability to observe something directly is extremely limited. Technological innovation has indeed stretched the boundaries of this limitation, but it has also led to a blurring of the principle of 'hard observation'. This blurring implies that the interpretation of the observation becomes more and more important, which in turn implies that philosophical and metaphysical constructions have become necessary for the development of scientific theories. The American philosopher and scientist George D. Gale expressed this as follows, in a poetic manner:

"We are currently entering a phase of scientific activities where the physicist has surpassed his philosophical foundation and where he, deprived of a supply of conceptual ideas, is ready and waits for a little help from his philosophical brothers in arms." [57-2]

This is a conclusion that is also shared by many other scientists and philosophers. A good example of the challenge of dealing with such great complexity in the development of scientific theories can be seen in string theory. String theory is basically an extension and refinement of quantum theory and is based on the idea that the elementary materials of matter are extremely small, miniscule strings that constantly vibrate. These vibrations occur at different frequencies and constitute the base of matter and energy. According to string theory these strings are considerably smaller than even subatomic particles. The physicist Michael Green explains that if an atom were to be blown up to the size of our solar system, about 10 billion kilometres, then one single string would be the size of a tree! String theory was developed in the 1970's as an attempt to unify the four fundamental forces, and therefore also to unify relativity and quantum physics.

This attempt led to five theories that were later united in M-Theory that is also called the Theory of Everything. String theory has a reputation for being extremely complex, and indeed, it is. Different versions of the theory, for example, assume an existence of a 10- or 11-dimensional reality! It is impossible to imagine a 10- or 11-dimensional reality, after all, just trying to imagine a 4-dimensional space-time continuum is already difficult enough. String theory is basically a very complex mathematical model; it is theoretical by nature and cannot really be visualized. Furthermore, there is no experimental evidence for the theory. The forces necessary to make the miniscule strings 'visible' in a particle accelerator are simply too big for the current equipment. Nevertheless, string theory has many adherents amongst scientists and, by the way, the basic thought of string theory is not fundamentally different from, for example, the quantum model, in which matter is also described as a state of energy. This state of energy is dynamic, and thus the image of vibrating strings or rubber bands is quite appropriate.

It is inevitable, although some scientists may not want to accept it, but at this point scientific explanations take on a philosophical and metaphysical form. The description of string theory illustrates this and is also a good example of it. The current generations of renowned scientists advance one fantastic theory after another in order to try to come up with an ultimate and all-encompassing explanation. The word fantastic is not used sarcastically here, but it refers to the exceptional levels of intelligence, innovation and extraordinary quality that have come to typify modern scientific theories. Just open any book about quantum physics, relativity, string theory, inflation theory or the big bang cosmology and you will be faced with the most amazing descriptions of reality. Reality is so complex, and science is so advanced in being able to analyse reality that we reach the limits of language and even the limits of our imagination. Words and concepts can be very rich, nuanced and even innovative, but they still have their limitations when making this complex reality understandable and clear. As the famous philosopher Wittgenstein formulated in his 'Tractatus Logico-Philosphicus':

"Whereof one cannot speak, thereof one must be silent." [63]

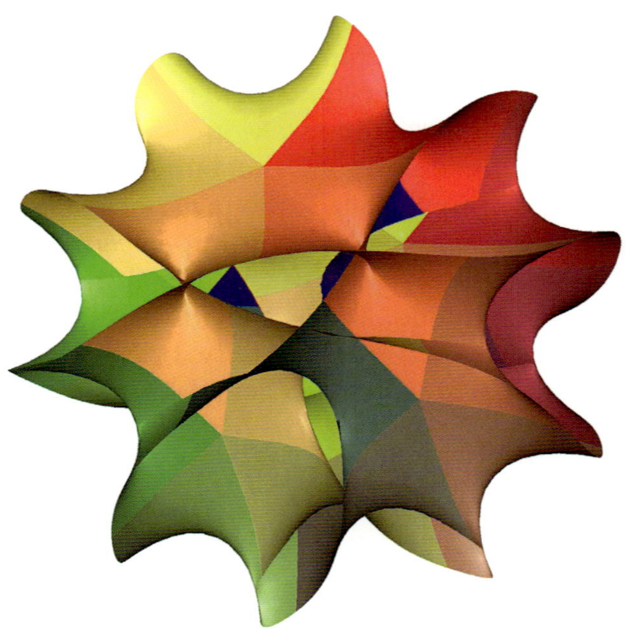

Image of a Calabi-Yaurui space, a very small and invisible dimension that is part of string theory.

This quote is one of Wittgenstein's famous quotes and refers to the limitations of language and the limitations of our imagination. In this statement, he basically rejects philosophical speculation and metaphysical doctrines for being non-verifiable and mostly incomprehensible and supra-rational. The problem is that the new physics and many recent scientific observations and experiments fit exactly into this category. They are barely verifiable, mostly incomprehensible, abstract and paradoxical. Based on the above-mentioned quote of Wittgenstein many scientists should therefore be silent, but this of course would not further the development of new and better scientific insights. Another problem with Wittgenstein's quote is that he contradicts himself. The quote is itself paradoxical: it refers to matters that we cannot speak of, and as a consequence Wittgenstein too should have been silent on them.

Nevertheless, it is clear that reality is unimaginably complex at all levels and that it has extraordinary properties. These extraordinary properties are not limited to the supernatural, the transcendent or the metaphysical, but they also manifest themselves in our visible, everyday world of matter. If we were to follow Wittgenstein's advice then we would not be able to discuss anything because, by definition, the language and thoughts used to describe or help visualize any form of reality—including everyday reality—are incomplete and imperfect. Whilst we will come back to this topic in more detail later in this chapter, the key point is that language, words and thoughts only refer to reality, and are representations of reality. If the representations are accurate, then this can lead to actual knowledge. Consider, for example, a map with which we navigate our journey through an unfamiliar landscape. The map differs quantitatively from the actual landscape. The map is only 80 x 80 centimetres and the landscape itself is tens or even hundreds or thousands of kilometres in size. However, from a qualitative point of view, a good map will provide a true image of the actual landscape and will therefore help the traveller to navigate the landscape correctly. Knowledge works in a similar manner, but only if it is accurate proper and correct knowledge.

Whilst Wittgenstein correctly stated that reality is very mysterious and often incomprehensible, we should add the qualifier that reality is 'understandably incomprehensible'. As Albert Einstein so beautifully puts it:

"The most incomprehensible thing about the universe is that it is comprehensible at all."

Wittgenstein's conclusion that we should be silent about matters that are too complex and incomprehensible to properly describe, is however, with all due respect for this great thinker, incorrect. We can indeed acknowledge that our consciousness and the thoughts, words and sentences arising from it are quantitatively limited. We can also acknowledge that the paradoxical nature of reality and the infinity of space, time and energy are beyond human intelligence. This, however, does not imply that our elementary tools, our consciousness, our thoughts and their expressions in the form of language are fundamentally insufficient. Reality exists, we are aware of reality, and since our consciousness is part of that reality, it can at least be comprehended qualitatively.

Therefore, pushing the boundaries of our perception has many consequences. Reality that looks superficially somewhat understandable, turns out to be dramatically different beyond this superficial appearance. As we dig deeper and deeper with the assistance of all kinds of amazing technological tools and aids, we are discovering that beyond this superficial appearance there is a hidden reality of an unimaginable complexity, an unimaginable size and that also has unimaginable qualities. Beyond this superficial appearance, reality appears to be infinitely 'elastic'. The great can still be greater, the miniscule can still be smaller, the full can still be fuller, and the complex can always be more complex. Human consciousness that tries to understand all this is seriously tested and is faced with its limitations time and time again. Fortunately, we are part of this very mysterious reality and therefore we are at least qualitatively closely connected to it. Because of this connection, the universe, in the end, remains understandable, at least in a qualitative way. This can be compared to a drop of seawater taken from the ocean; qualitatively they are made of the same substance and share the same properties even though they are quantitatively incomparable.

The consequence of pushing the boundaries of our perception is to realize that everyday reality is actually extraordinary and that extraordinary reality is exceptionally extraordinary. In a certain way this makes the extraordinary actually quite ordinary again, something we will discuss in the following section.

The extraordinary reality – energy

What is reality?

The second problem that materialism is faced with is that what scientists ultimately observe is beyond human imagination. The descriptions of the subatomic universe, as explained above, cannot be grasped and are absolutely extraordinary. We can only try to visualize this reality by using the examples and illustrations used in some of the previous discussions. However, we cannot fully understand it, and as a consequence, we must conclude that even the common, ordinary reality of hard, touchable matter is in fact itself extraordinary. This analysis is directly linked to the discussion about atheism versus theism and the second proposition of atheism. What this discussion shows is that the traditional differences between the observable physical reality and the so-called supernatural or non-material reality have largely faded. Many of the arguments used in the atheism versus theism discussion, are focused on this issue. We again refer to the previously quoted Richard Dawkins in his book *The God Delusion* where he gives his definition of atheism:

"An atheist in this sense of philosophical naturalist is somebody who believes there is nothing beyond the natural, physical world, no supernatural creative intelligence lurking behind the observable universe, no soul that outlasts the body and no miracles — except in the sense of natural phenomena that we don't yet understand. If there is something that appears to lie beyond the natural world as it is now imperfectly understood,

This definition effectively exposes one of the inherent weaknesses of the atheist point of view, which is the so-called distinction between the physical world and the non-physical world. We are used to dividing the world into ordinary versus extraordinary, natural versus supernatural, common sense versus incomprehensible. With the rise of science hundreds of years ago, one of the strongest arguments in support of the scientific method was that the supernatural could now be explained by getting a better understanding of the natural. What seemed to be extraordinary turned out to be an incomplete understanding of the ordinary. In other words, the extraordinary, the supernatural and the incomprehensible were just examples of ignorance and a lack of understanding of the 'ordinary' natural world. Science was down-to-earth and based on convincing empirical observations and solid facts. Objects were studied, touched, smelled and heard. What was even more important was that anyone who was willing to make an effort could make these same observations. Therefore, science was concrete, precise, and verifiable by all. Individual observations of certain phenomena could be linked to other observations by means of mathematics, for mathematics was also a very exact science. Natural phenomena turned out to be structured and predictable and were elevated to the status of a natural law, expressible in concise mathematical equations. This enabled scientists to predict and manipulate future events giving birth to applied science and technology.

Technology, as we now know it, has influenced and changed the world more than anything else that we have experienced in recent history. Technology was the force behind the Industrial Revolution from which our economic circumstances improved considerably, medical science and health care progressed, and the duration of our lives has increased. On the negative side, it also brought technological mass warfare and weapons of mass destruction. But nevertheless, because of these impressive results science became most credible. The scientific method became the standard in the search for truth and scientists became the high priests and authorities of a new church, the 'scientific church'.

This is not meant sarcastically, for there is no doubt that the scientific method and scientific research are very valuable. This also holds for the practical applications of scientific knowledge in the shape of technology. However, we also have to acknowledge that science, at a certain moment, had to compromise its own principles, namely the principle that the foundation of knowledge must be empirical hard, observable facts. Science has gone through a major change, especially in the 20th and 21st century. Apart from this compromise with empiricism, the distinction between the ordinary and the extraordinary has also faded. Science once started out by making the extraordinary ordinary, but in the end, it has itself become extraordinary. Or, in other words, due to science the ordinary world itself now became extraordinary, but another kind of extraordinariness then the ignorant and superstitious type of the Middle Ages.

The discovery of force fields

The unravelling of the materialistic world view and the world of the 'ordinary' actually already started in the 17th century, when Sir Isaac Newton first discovered gravity. Whilst all scientists today accept gravity as a reality, the mechanism of gravity remains a big mystery for the mechanism itself has never been observed. In this respect it is absolutely remarkable that the existence of such force fields and energy fields have come to be accepted as a matter-of-course, both in scientific circles as well as in popular scientific explanations — and when we say as a matter-of-course we mean that they have come to be considered as integral parts of the material universe and an obvious extensions of matter. This is a truly remarkable assumption given their entirely mysterious nature, and upon closer examination, it is also incorrect. Newton's discoveries in the area of mechanics were mainly focused on the interactions of matter that were based on physical contact, discoveries that he could understand and explain. However, his discovery of the concept of 'force', as it was related to gravity, was not at all straightforward and obvious. He described gravity as something that could influence a massive object across a spatial distance, even in a vacuous space. Yet, he was not happy with this definition, or with the notion of 'action at a distance' as suggested by his equations. Newton wrote in 1692 in his third letter to the philosopher Richard Bentley:

"That gravity should be innate, inherent and essential to matter, so that one body may act upon another at a distance through a vacuum, without the mediation of anything else, by and through which their action and force may be conveyed from one to another, is to me to great an absurdity, that I believe no man who has in philosophical matters a competent faculty, can ever fall into it."

In his General Scholium of 1713, in the second publication of *Principia*, he stated:

> *"Thus far I have explained the phenomena of the heavens and of our sea by the force of gravity, but I have not yet assigned a cause to gravity. And hypotheses, whether metaphysical or physical, or based on occult qualities, or mechanical, have no place in experimental philosophy. It is enough that gravity really exists and acts according to the laws that we have set forth and is sufficient to explain all the motions of the heavenly bodies and of our sea"* [58]

Newton was very uncomfortable with the mysterious nature of gravity's mechanism, but he accepted it as a fact. He did not try to gain further insight into it or to explain the mechanism. The mathematical equations with which he described the functioning of gravity were working fine and this was enough for him. Furthermore, gravity was such a part of everyday life that people were not even aware of it until Newton pointed out its existence. Prior to Newton, it appeared to be a matter of common sense that an object is heavy because it has mass, and therefore, it falls towards the surface of the earth. Newton, however, discovered that an object is not heavy because it has mass, but because gravity draws one mass to another. We now know that objects in space are weightless and there is no above or below in space. Earth itself, after all, is a very heavy, massive object that floats through space. It actually circles in a weightless state around the sun, whereby the centrifugal power of its orbit around the sun and the gravity of the sun remain perfectly balanced.

The idea of a force within the scope of the laws of nature became even more important in the 19th century when scientists discovered that there are more forces in the universe than just gravity; these were electricity and magnetism. Mankind was aware since antiquity that these forces existed, but did not understand them. In the 19th century several scientists carried out research in this area, of whom the work of James Clerk Maxwell became most prominent. Maxwell laid the foundation for the science of electromagnetism.

He unified the work of other scientists and presented a number of mathematical equations that eventual became known as the laws of Maxwell. Maxwell, in fact, demonstrated that light, electricity, and magnetism are all part of the same field force, electromagnetism, and that they were just different manifestations of the same energy field. Maxwell, however, still believed that a 'force' was based on some form of physical contact. Similar in a way that sound waves are caused by waves propagating through a material substance, the air, he assumed that light waves were propagating through an even more subtle, yet material substance, the luminiferous aether.

At the beginning of the 20th century, however, it became apparent from various scientific experiments, notably the Michelson-Morley experiment, that the notion of the luminiferous aether had to be abandoned. From then on it was agreed that electromagnetic waves, such as light, are able to propagate themselves in vacuous space. Forces, therefore, were able to exist outside of material particles and physical objects could be influenced from a distance without physical contact. Thus the term 'force' took on a whole new meaning, and by implication caused to undermine the classical mechanistic image of the universe. The central aspect of mechanistic thinking was that an object could only influence another object by means of direct physical contact. This is the effective meaning of the word 'mechanism' or 'mechanistic'. This idea is undermined by the existence of forces, for by means of force one object can influence another object from a distance without physical contact. On one level we do not know what a force really is, for it is invisible, it has no mass and it is not bound to a particular place. We can only indirectly observe a force by the influence it exercises on the visible objects around it. This implies two things:

1. *Matter is not the only substance in the universe, but there is also a non-physical reality, force fields, which can also be described as energy fields.*

2. *The strictly mechanistic image is no longer valid. Objects can influence each other without physical contact.*

Therefore, the discovery of force fields or energy fields is a serious impediment for mechanistic materialism, for it appears to be incorrect, or at least incomplete. While the existence of the luminiferous aether was abandoned later discoveries, particularly in the field of quantum physics, led to the notion that so-called 'vacuous space' was not in fact empty. It turned out that the discoveries of the sub-atomic reality in combination with the discovery of energy fields, led scientists to the conclusion that space itself is in fact a massive energy field, called the 'quantum vacuum' state full of 'vacuum energy'. The 19th century luminous ether was effectively replaced by the infinitely more subtle vacuum energy. This remarkable development will be discussed in much more detail in later sections.

Einstein's theory of relativity

The further unravelling of materialism was precipitated by Albert Einstein's introduction of the special and general theory of relativity in the beginning of the 20th century. The special theory

Albert Einstein

James Clerk Maxwell

of relativity had a dramatic impact on scientific thought. Einstein introduced his theory in 1905, and it was followed by the general theory of relativity in 1916. Relativity triggered a revolution in science for various reasons. One of the biggest consequences of the theory of relativity was that it directly challenged many of the ideas and theories of reality and the universe that had been based on common sense. For example, Einstein introduced the notion that the speed of light is a constant and is not affected by the motion of the observer. Say we fly in a spaceship away from earth at half the speed of light, or at 150,000 km/sec. If we were to then measure the speed of a ray of light that is moving past our spaceship we would expect to obtain a measurement of 300,000 km/sec (being the speed at which light travels) minus our own speed of 150,000 km/sec = 150,000 km/sec. However, contrary to all common sense, our measurement would result in something completely different. To our astonishment, we would measure that the ray of light is still travelling at 300,000 km/sec and not the expected 150,000 km/sec. Einstein explains this by stating that at such speeds something happens to the connection between time, space, and moving mass. Mass expands, and time slows down, so the ray of light that passes our spaceship still moves past us at a constant speed of 300,000 km/sec even in relation to our own speed. In fact, Einstein stated that time, space, mass and energy are all interconnected. If the value of one of these elements is changed, then the values of the other elements also change in the same proportion. This is one of the cornerstone principles of relativity.

Another example of such an extraordinary phenomenon is the connection between mass and energy. Einstein expressed this relationship in his world-famous equation $E=mc^2$. This equation also defies normal human comprehension. Before Einstein's discovery, scientists knew that mass could transform into energy, for example, by burning a block of wood or another flammable object and thereby transforming it from mass to heat, light, and some ashes. In scientific terminology this process is called the combustion of chemical energy and was already clearly understood in the 19th century. However, a major concern in those days was the life span of the sun. Scientists were wondering how long the sun could keep generating so much heat before its fuel would run out. Scientists assumed at the time that the energy of the sun was caused by chemical combustion and that the sun continued to exist because it contained an enormous amount of flammable material to be used as fuel. The physicist Lord Kelvin estimated that the life span of the sun would be no more than 10 million years and made the following statement:

"The inhabitants of earth cannot continue to enjoy the light and heat essential to their life, for many millions of years

longer, unless sources now unknown to us are prepared in the great storehouse of creation." [65]

Einstein revealed these sources when he laid the foundation for the discovery of nuclear energy. Einstein knew that atoms were not indivisible, but consisted of other, more basic particles that were being held together by especially powerful forces that contained massive amounts of energy. While Ernst Rutherford had indicated that atoms consisted of smaller subatomic particles, Einstein was able to state more fundamentally what the connection was between the structure and the mass of the atom and its energy contents. The revelation that truly shocked everyone, however, was the enormous amount of energy contained in just a single atom. Based on these discoveries, we now know that the sun is driven by a process of nuclear fusion and will continue to burn for another 5 billion years. These discoveries also led to the development of nuclear weapons, the atomic bomb and the hydrogen bomb, which are all weapons with an incredible destructive force. The bomb dropped on Hiroshima on 6 August 1945, for example, had a mass of only 60 kilogrammes of uranium, but created an explosion equivalent to 15,000 tons of TNT. Two-thirds of the city area was destroyed, and 100,000 of the estimated 350,000 inhabitants of the city died instantly. By the end of 1945 the number of dead had risen to 140,000. The power that the atom contains is truly extraordinary. J. Robert Oppenheimer, the nuclear scientist who led the Manhattan Project which developed the first nuclear weapon, was absolutely astounded when he saw the first atom bomb explosion in the desert of New Mexico on 16 July 1945. He quoted a line from the Bhagavad-Gita, the book that contains the essence of the Vedic philosophy of India:

"Now I am death, the destroyer of worlds"

(Bhagavad-gita, chapter 11, verse 32)

Whether it is for better or for worse, the discovery of nuclear energy has changed the world forever. More than with any other scientific discovery, humans have been faced with the extraordinary, mysterious, and almost magical nature of our universe. What Einstein made especially clear was that force fields are the same as energy fields and that matter is a

The first atomic bomb, used for war purposes, was exploded over Hiroshima. The bomb 'Little Boy' had a power of 15.000 tons of TNT.

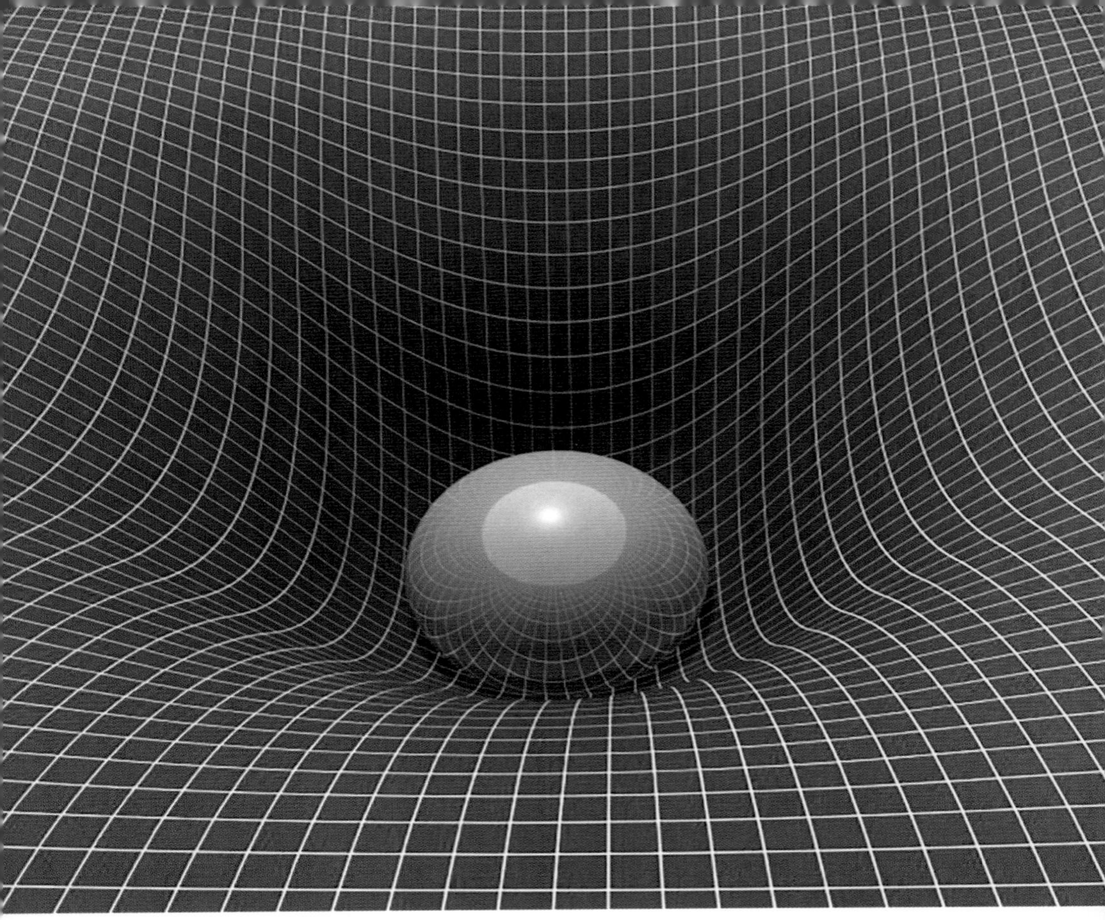

Illustration of Einstein's explanation on the functioning of gravity. The ball represents a planet that, because of the size of its mass, deforms the space around it. This deformation of space generates a counterforce that we experience as gravity. Even though this explanation is rather esoteric—for how can an 'empty' space be deformed—this explanation is regarded as the most plausible explanation by most scientists.

manifestation and transformation of energy. Gravity, something that Newton could not grasp, was described by Einstein as one of the many energy fields in the universe, in the form of the time-space continuum. According to Einstein this energy field is deformed because of the presence of mass, a planet for example. This deformation generates a certain counterforce that we experience as gravity. When this is explained in physics books for students, we can usually see a nice picture of some sort of heavy metallic ball that lies on a mattress-like surface, partially sunken because of its weight.

The mattress, representing the time-space continuum, below the ball is deformed and is pushed downwards, bringing an upward counterforce to bear on the ball. That counterforce is what we experience as gravity. It is important to note, however, that this is only possible when there is something in space. A totally empty space does not have the ability to bend and therefore cannot create any counterforce. A space made of energy, on the other hand, does have this ability. Admittedly, this is a rather abstract description of the tangible reality that gravity is. The remarkable thing, however, is that energy is described by Einstein as a concrete reality even though it totally differs from the material reality as we experience it. Energy does not have mass and is not localised. Matter, on the other hand, does have mass and is localised. Generally, we are not able to perceive energy directly, but we do perceive matter directly. The materialistic worldview is unable and inadequate in trying to explain the mysterious nature of forces and energy. While the initial discoveries of the existence of force fields had led to the notion that forces were a by-product of matter, currently, the tables have completely turned. For it is now clear that matter is effectively a by-product of energy. Energy and forces form the foundation of the atom and material mass is but a transformation of that energy. Einstein's formula $E=mc^2$ has demonstrated this fact and placed energy centre stage in our understanding of the universe. Energy is the base of gravity and all other forces in the universe, and even space itself is energy. Energy is also the foundation of the intelligent cohesion and complexity that we perceive everywhere in the universe. Energy exists, but it behaves completely different from material reality. While energy is the base and foundation of matter, it is itself non-material which fact completely undermines materialism. Then what is energy precisely?

What is energy?

When we think about energy in everyday life, we first of all think of activity and motion and the ability to perform work.

Thereby energy can take on many different forms, such as motion, heat, pressure, gravity, electricity, etc. Furthermore, one type of energy can be transformed into another type of energy. Electrical energy, for example, can be transformed into heat and vice versa. Whatever form it takes, in the end, work can be performed with it. Energy in this sense has everything to do with motion and dynamics. Another important fact is that energy can indeed be transformed from one form into another, but it is never lost. The law of conservation of energy is a fundamental law of nature. Einstein's theory of relativity shed a completely new light on energy. Einstein stated in his formula $E=mc2$ that mass is also a form of energy. Of course, mass is a substance and that is how we experience it. But more fundamentally, it appears

What is energy? We talk about it constantly, we use it every day, but fundamentally we do not understand it. Energy is one of the most fundamental mysteries of the universe. Energy is the foundation of existence, it is the cause of matter or mass, but it is more than both of them. Energy is the reservoir of all properties, and it possesses all these properties, including even the opposites of all these properties, simultaneously. Energy is the cause of all change, but it does not change itself. Furthermore, energy consists of pure intelligence and consciousness. In later chapters thorough attention will be paid to the very mysterious properties of energy.

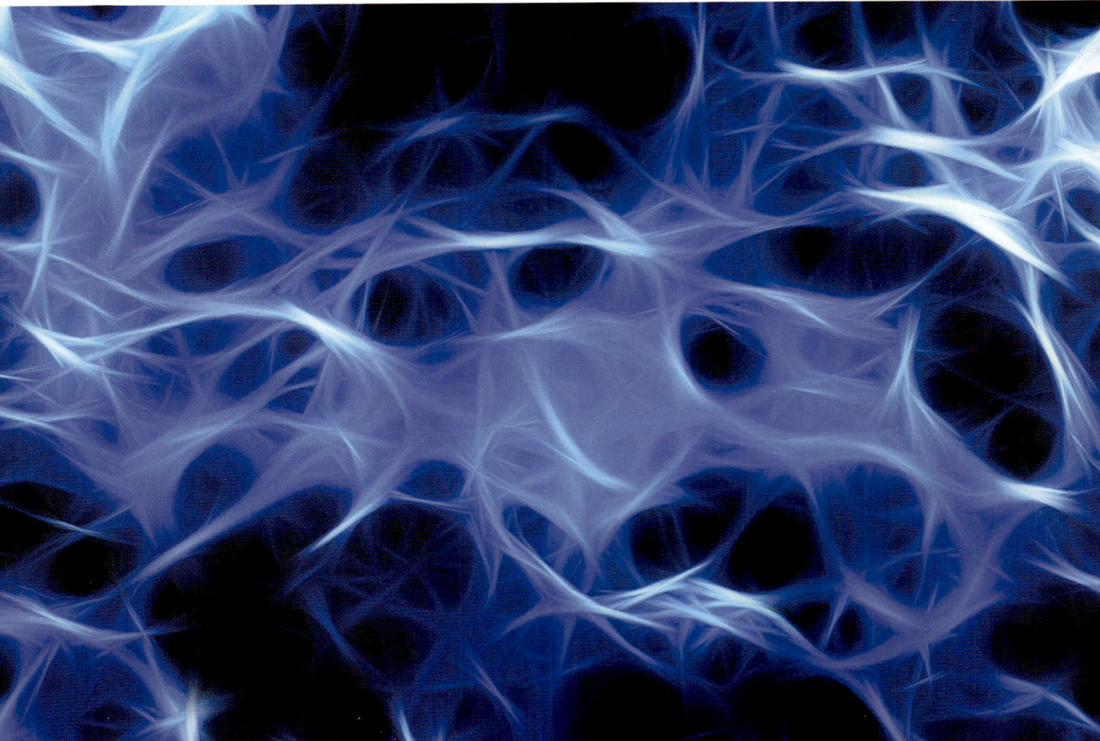

that mass itself is made of energy. This is the core of the theory of relativity and also quantum physics. The atom, or the sub-atom is, in the end, a state of energy, which implies that energy is more than just motion or the ability to perform work. $E=mc2$ has promoted energy from being merely something that causes motion to something that is both the cause of motion as well as the substance that is actually being moved; a truly astonishing phenomenon.

There is yet another mysterious aspect of energy which has to do with the fact that energy is a field and is not localised. By nature of being a field, and given that within that field it is omnipresent, the normal laws of space and time are not applicable. The omnipresent nature of an energy field implies that each part of that field is connected simultaneously with every other part of that field. This has led to the observation of a phenomenon that we call synchronicity. Usually time is necessary to cover a distance from A to B, regardless of the speed with which the object moves from A to B. If A and B, however, are part of a field, then no distance at all needs to be traversed. Moreover, in that case A is identical to B and B is identical to A, if something happens to A then this will happen simultaneously to B and vice versa, without time or distance coming into play.

Since the universe is filled with energy fields and is itself actually an infinite field of energy, the laws of space and time operate differently. These laws would be characterized as super natural, since the limitations of space and time are no longer applicable. In quantum physics scientists have observed phenomena that indicate such a non-local connection of parts of an energy field. A good example of this is what scientists call entanglement, where different particles are situated at a great distance from each other, yet appear to be connected. This connection manifests itself because a change in one particle causes a similar change in another particle. These changes occur simultaneously over great distances. Many experiments have been conducted that have confirmed these phenomena. Einstein called this phenomenon 'spooky action at a distance' and he was quite startled by it.

However, we have actually observed such a 'spooky action at a distance' phenomenon before, for example when Newton discovered the law of gravity. Force fields are exactly that; they generate simultaneous action over a distance. The explanation for this is that the field is a unity and each part of the field is connected to every other part of the field. For objects that are situated within a field there is no distance, but for objects that are situated outside the field and that are not part of the field, distance becomes a distinct reality. While we observe this phenomenon, it is hard to comprehend rationally, yet, on the other hand, it is completely logical. The logic is that, in the end, each substance or energy field has a certain size, no matter how big or small it is. Within that size, it is a unity where each part is connected to each other part. In a way, we have trouble grasping the concept of unity, because unity is so obvious and self-evident and so intertwined with existence itself, that it is hard to treat it and analyse it as if it were a distinct part of reality. All objects that we perceive, we perceive because they distinctly exist and they exist because they are distinctly one. And since we do not have an alternative for existence, other than non-existence which does not exist, we experience existence as self-evident, and incomparable.

Since comparison is a key ingredient in acquiring knowledge and understanding, a lack of comparison makes things more difficult to understand. Therefore it is sometimes difficult to grasp the various intrinsic attributes of existence, such as unity, substance, properties and energy. Energy, in particular, challenges our imagination: it is everywhere but at the same time it is intangible; it is the core of reality and it is the foundation of substances, but at the same time it is above and separate from all substances—it is a fundamental unity that simultaneously contains an infinite multitude. The special feature of a unity is that it is fundamentally simple, for after all it is only one and almost nothing could be simpler than one; the only thing simpler than one is zero or nothing. However, nothing does not exist; as has been remarked before. On the other hand, this fundamental and simple unity is simultaneously infinitely complex, since unity cannot exist without a multitude. Imagine for example the existence of a totally indivisible object, it must still consist of parts, if only because of the fact that it has a size in space and within that size it consists of spatial parts. Space can be regarded, by definition, as a unity and multitude simultaneously, just like Euclid defined a line as a set of infinite small dots. The line is a unity, but at the same time it is a set of an infinite number, and

Zeno of Elea

hence a plurality, of small dots. Fundamentally this is a paradox, a paradox on which, for example, the comic contradictions of the Greek philosopher Zeno are based on.

The paradox of Achilles and the tortoise is particularly famous. Achilles, a professional runner and warrior, is challenged by a tortoise to a race. The tortoise states that as the challenger he will have a head start of 100 metres and that Achilles will never be able to catch up with him, no matter how fast he runs. Achilles starts laughing out loud. Then the tortoise explains why he is going to win. He asks Achilles how fast he can cover the 100 metres. After Achilles has explained how much time he will take, the tortoise then explains to him that during this time he will have covered ten metres himself. The tortoise continues his argument and states that whilst Achilles tries to cover those ten metres, the tortoise will also advance a small distance and will still be ahead. Each time that Achilles covers a certain distance, the tortoise will have also moved on a bit. This will continue infinitely. Even though the differences keep getting smaller, the tortoise argues, Achilles will never be able to catch up with him. After this explanation, Achilles is so totally discouraged that he decided not to run the race and admits that he has lost.

Whilst the tortoise's argument is logical and compelling it completely defies our experiences and common sense. The answer is largely determined by how we define and interpret the challenge. On one hand, if we define a line as a set consisting of an infinite number of infinitely small points, then we would have to agree with the argument of the tortoise. However, if we define the line as a unity and a smooth continuum, consisting of a finite measurement, then of course Achilles can catch up to the tortoise and overtake him without any problem. Zeno gives another example in which he states that an arrow (shot from a bow) can never move. Zeno states that the arrow either moves at the location where it is, or at the location where it is not. If it moves at the location where it is, then it cannot move, for it is there at that specific location. If it were to move at the location where it is not, then it couldn't move either, because it is not there. Zeno concluded therefore that an arrow can never move;

moreover, he concluded that movement was not possible at all. Again, common sense and our direct experiences would disagree with this. We are again faced with the same paradox, namely the paradox of the definitions of space and time. Are space and time a set of infinite small units, or are they finite continuums—in other words, unities—in which parts do indeed exist, but are primarily and fundamentally one? If one chooses for the infinite set of particles, then distances and time units could not be covered if one would present them in a mathematical form. Fundamentally time and space are simultaneously one and many, whereby they manifest themselves both as continuums and as an infinite set of infinitesimal units. While they are one and many there is a hierarchical relationship between them, the many is always part of and in a way subordinate to oneness, for the many are encompassed by oneness. In practice and based on experience this hierarchy is also manifest; we experience a certain distance in space as a finite unit, encompassing and including the manyness.

Until the 17th century philosophers and mathematicians had a very difficult time unravelling these paradoxes. Only after the development of modern mathematical systems, such as calculus, were they able to solve the problem. The essence of the solution, from a mathematical point of view, was that the sum of an infinite amount of terms in a series can nevertheless produce a finite result. The infinite number of time spans that Achilles needs to reach the points passed by the tortoise can be added together, based on modern differential and integral calculations, to result in a finite total time: this is the time that Achilles needs to catch up to the tortoise. There is another amusing anecdote related to Zeno. The story goes that a student of Zeno was presenting his argument to a gathered crowd and using a lot of arm gestures while he spoke. In his presentation, the student explained that all the movement he was making was actually just an illusion and gave various examples to support his case. Suddenly, however, whilst he was gesticulating quite strongly, the student dislocated his arm. He was in great pain and called for help from a physician. Someone from the audience came forward and started explaining, however, that it was impossible

Heraclitus

that he could have dislocated his arm. Either, he had dislocated his arm at the place where his arm was located, but then it could not be dislocated there. Or it was dislocated at the place where it was not located, but since his arm was not present there, it could not be dislocated there either. Nevertheless, the student still demanded a physician, for he was in great pain, regardless of the philosophical implications.

There is, however, a deeper and real paradox on which the paradoxes of Zeno are based. Zeno was a follower of the Greek philosopher Parmenides, who considered 'being' and the fact that things are unchanging as the foundation of reality. According to Zeno and Parmenides only 'being' was real and change was just an illusion. For this reason, Zeno used his paradoxes to indicate that movement and change is just an illusion. The views of the Greek philosopher Heraclitus' were exactly the opposite of those of Parmenides. Heraclitus stated that only change is real and that 'being' and a state of permanence are illusions. Heraclitus made the famous statement that one can never step into the same river twice. The notions of permanence versus change are contradictions that have been on the minds of philosophers for thousands of years.

Modern science has also been confronted with this paradox, particularly in quantum physics. As Capra explained previously, sub-atomic particles move at unimaginable speeds. The electron makes 3000 trillion revolutions per second around the nucleus of the atom. At the same time the atom itself remains stable and interacts with other atoms to create stable molecules. These stable molecules form the base of the material world and material objects as we experience them daily. These objects have mass and they have different gradations of hardness and they exist in relative stability, or at least we are not under the impression that there are moving parts in these objects doing 3000 trillion revolutions every second. This is a clear example of how movement and permanence exist simultaneously, and they do not exclude each other. Furthermore, in more philosophical terms, when there is movement there has to be 'something' that moves and 'something' that remains permanent to a certain extent. It is not movement itself that is moving, it is not change itself that is changing, it is something else, something permanent that is moving and changing. Movement and permanence cannot be seen separately, despite the fact that they relate to each other in a paradoxical way.

Indeed, quantum physics contains a number of paradoxes. Previously we discussed how sub-atomic particles sometimes behave like a wave—or in other words, an energy fields — and sometimes as a particle. In this case, we also observe a paradoxical contradiction of a particle (multitude) and a field (unity). Other comparable paradoxes include the relation between energy and mass, between the organism and its components, between substance and its properties, between permanence and change.

It seems that reality is fundamentally organized this way and that reality itself is actually a supra-rational paradox. This supra-rational paradox is the core and the cornerstone of the 'ultimate reality'. This ultimate reality is made of energy and in later chapters we will explain the attributes of energy in more detail. For now, we can conclude that energy is the foundation of reality, an unbounded field of existence that is the reservoir of all actual and potential properties that are simultaneously present there. Energy is fundamentally paradoxical and contains some, ordinarily, irreconcilable contradictions. The source of these paradoxical contradictions is the simultaneous unity and multitude of all things.

Furthermore, the unrelenting conclusion from the teleological argument and the anthropic principle as explained in the previous chapters is that energy is conscious and intelligent. It is this infinite intelligent unity that organizes and structures the infinite multitude in an infinitely complex and intelligent way.

The non-phenomenal world – dark matter and dark energy

The third problem encountered by materialism was the fact that scientists were forced to propose the existence of realities that are entirely unobservable. As stated previously, this was already a problem since the 17th century when Isaac Newton discovered the phenomenon of gravity. He accepted the existence of this force by the observable effect it had on material objects, but he could not directly observe or explain the force, or rather, he could not explain what exactly was the 'substance' of this force. The discovery of forces in science, starting with gravity and later on electromagnetism and the nuclear forces, presented indeed a major problem for the empirical foundation of science. The force itself was not observable, only its effect on material objects was; which is, of course, still a form of observation, even if it is indirect observation.

What has been even more problematic, however, was the 'discovery' of dark matter and some decades later the discovery of dark energy. The discovery of dark matter occurred while studying star systems and the behaviour of matter in these systems. It appeared from their behaviour that the amount of matter that could be seen in these star systems was insufficient to generate the gravity needed to keep them together. Therefore, there had to be an invisible substance that could also generate gravity, but which was significantly different from the 'normal' properties of matter. This led to the formulation of the existence of 'dark matter'.

Dark energy was discovered following recent measurements which indicated that the expansion of the universe is accelerating instead of slowing down. This expansion is attributed to the existence of an energy that is not visible or observable in any way, and therefore it has been called dark energy and is assumed to have an opposite function to gravity. Previously, it had been assumed that the expansion of the universe that followed the big bang would slow down in due course of time because of gravity. In the end, it was assumed, the expansion would come to a standstill and that gravity would then cause it to move in the opposite direction. The universe would implode, resulting in what was called the big crunch, at which point the universe would return to the state of the singularity. However, the recent observations that the expansion of the universe is accelerating has led scientists to theorize that there must be an invisible force at work, stronger than gravity, which was then coined for want of better word, 'dark energy'. Yet neither dark matter nor dark energy have ever been directly observed. Furthermore, scientists have come to the shocking realization that dark matter and dark energy are probably responsible for 96% of the contents of the universe. In other words, the material universe that is visible to us only represents 4% of the total amount of matter and energy in the universe!

For astrophysics and cosmology dark matter is an unknown composition that does not mirror or radiate enough electromagnetic radiation to make it possible to observe it directly. We know, however, that it exists, thanks to the influence of gravity on observable matter. In order to fit into the standard

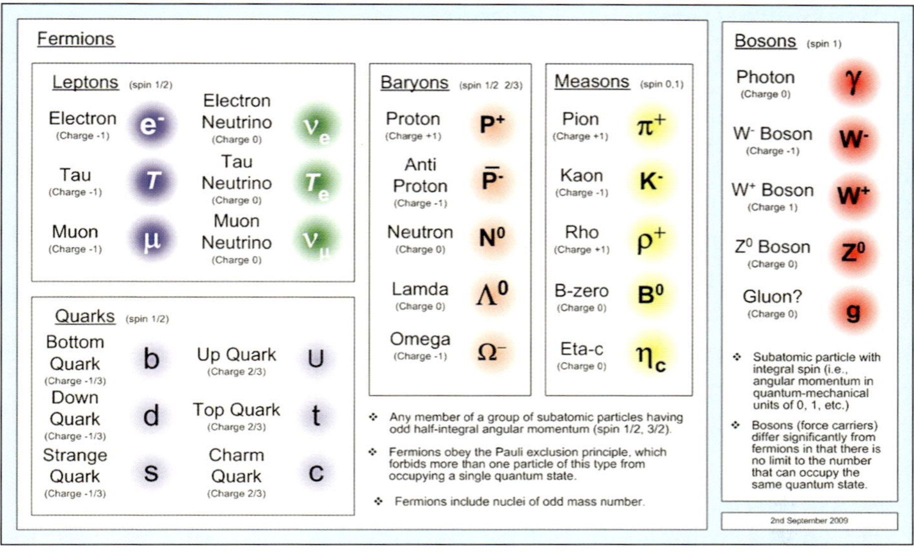

Above is an overview of the standard model, consisting of fermions, quarks, leptons, bosons and their various sub-divisions. In total 61 sub-atomic particles have been identified in particles physics that are shown in the scheme above. Even though a lot of progress has been made regarding the research into the fundamental building blocks of matter, there is still no all-embracing theory in which all particles and forces are combined. Therefore, the standard model is also called the 'Theory of Nearly Everything'.

model and the current Big Bang cosmology, it is necessary that this matter has energy and mass, but also, that it is not made from common, or what are called baryonic atoms. The prevailing point of view is that most dark matter is non-baryonic. Non-baryonic matter consists of elementary particles that are not part of the standard model. The standard model consists of particles that are known to us such as electrons, protons, neutrons, but that also include a whole range of new, more recently discovered and less familiar subatomic particles. The family of sub-atomic particles has expanded vastly over the last decades, with, amongst other things, neutrinos, variations of quarks, fermions, leptons, photons, gluons, bosons and variations of the Higgs particle. To this, some so-called anti-particles can be added, such as the positron and the anti-neutrinos.

Some scientists have suggested that dark matter consists of a yet to be discovered elementary particle. These could be super symmetrical particles, such as axions, sterile neutrinos and WIMPs, Weakly Interacting Massive Particles. However, these hypothetical particles have not yet been observed in a laboratory. These are not particles that are part of the standard model, but particles with very high energy which formed in the beginning phase of the universe and are still floating around. Dark matter is therefore often associated with super symmetrical particles. Super symmetry refers to particles that have an anti-particle of exactly opposite properties. In many super-symmetry models, possible stable dark matter occurs in the form of the Lightest Supersymmetric Particle (LSP), of heavy, sterile neutrinos.

The Swiss astrophysicist Fritz Zwicky was the first scientist to point out the possible existence of invisible forms of matter and energy. In 1933 he observed the movements of star systems and came to the conclusion that their orbits did not coincide with the amount of gravity that could be generated from the visible mass. Other scientists complemented these results. Astronomer Vera Rubin presented a major discovery in 1975 at a meeting of the American Astronomical Society: most stars rotate in spiral-form star systems at about the same speed. This means that there was equality between their mass densities, much more than between the locations of most stars. She published an influential paper in 1980 where she explained these results. It appeared from this publication that either Newtonian gravity is not applicable everywhere, or that, according to a careful estimation, more than 50% of the mass of the star systems is contained in the relatively dark galactic halo. Rubin's results and conclusions were not warmly welcomed initially, but she stuck to her conviction that her observations were right. Other astronomers began to complement and continue her work, and they also concluded fairly quickly that most star systems are basically dominated by 'dark matter'.

What was even more staggering was that according to these scientific observations and calculations, only 4.6% of the matter of the universe consists of visible matter, made up of normal or baryonic atoms. The other 95.4% consists of 72% dark energy and 23% dark matter. In other words, more than 95% of matter

172

and energy in the universe is structurally and fundamentally invisible, unobservable and untraceable. Martin Rees writes in his book *Just Six Numbers*:

"The cumulative evidence for dark matter is now almost incontestable. The way stars and galaxies are moving suggests that something invisible must be exerting a gravitational pull on them. This is the same line of argument by which we infer the existence of a black hole when a star is seen to be orbiting around an invisible companion; it's also the reasoning used in the nineteenth century when the planet Neptune was inferred to exist because the orbit of Uranus was deviated by the pull of a more distant unseen object. [66]

Martin Rees acknowledges in his book that it is humiliating for science to not know what more than 95% what our universe consists of:

"It is embarrassing that more than ninety percent of the universe remains unaccounted for — even worse when we realize that the dark matter could be made up of entities with masses ranged from 10^{-33} grams (neutrinos) up to 10^{39} gm (heavy black holes), an uncertainty of more than seventy powers of ten." [67]

The existence of dark energy has been a more recent suggestion. Einstein did this by accident when he speculated about the existence of a 'cosmological constant' in 1917. Later on, Einstein called this hypothesis his biggest blunder, yet as we shall see it became the precursor to the discovery of dark energy. Michael Turner came up with the term 'dark energy' in 1998 and it strongly resembles the term 'dark matter' that Fritz Zwicky had come up with in the 1930s. The first immediate proof for dark energy arose from the supernova observations of the High-Z Supernova Research Team in 1998, followed by publications of the 'Supernova Cosmology Project' in 1999. These observations confirm that an acceleration took place with respect to the expansion of the universe, which was then later confirmed by several independent sources and observations. Since 1998, cosmological observations have become more and more rigorous. The most recent observation is the Supernova Legacy Survey (SNLS) of 2005. It appears from the first results of the SNLS that the average amount (meaning the comparable conditions) of dark energy agrees with the cosmological constant of Einstein with a precision of 10%. From a recent survey by the Higher-Z team of the Hubble Space Telescope, it appears that dark energy has been present for at least 9 billion years and even before the acceleration in the rate of cosmic expansion started. The discovery of cosmic background radiation, the assumption of the existence of the cosmological constant and the discovery of dark matter and energy have together led to what is called the Lambda-CDM model, or in other words the standard model for big bang cosmology. This standard model is now the generally accepted theory referring to the way the universe came into being. The existence of dark matter and dark energy has been accepted and included in this model.

This is the current reality of our knowledge of the universe. We have made enormous progress in science and technology, but we have reached the baffling conclusion that we do not know 95% of the universe. This is a humiliating realization that furthermore suggests that the mechanical, materialistic worldview is extremely incomplete. It is only a very partial understanding of the full reality. Even though gravity, dark energy and dark matter can only be perceived indirectly, the difference between the two of them is that gravity manifests itself in our immediate surroundings. We can 'feel' gravity and we can perceive the effect of it in our immediate surroundings. Dark matter and dark energy, however, are far more remote from our day to day experience and they can only be perceived by observations and measurements at the level of star systems and the macro structures of the cosmos.

Whereas at the beginning of the scientific revolution, scientists thought that the non-perceptible is an extension of the perceptible, currently the situation is reversed. It is now clear that the perceptible is an extension of the non-perceptible. This implies that the vastest part of reality is actually a non-perceptible reality. This also implies that we do not know what the properties are of the non-perceptible reality. They could be properties that look like material reality as we know it. However, they could also be properties that deviate strongly from what we know. In the end, we will have to determine this based on the properties of reality as they can be perceived by us. One thing is clear, however: the proposition that nothing exists outside the observable and tangible material reality is completely wrong and totally outdated.

The origin of matter and the origin of the universe

The fourth major problem for materialism was the discovery that the universe had an absolute starting point culminating in what is now referred to as big bang cosmology. During the first half of the 20th century, whilst significant research and advances were being made in our understanding of the structure of the atom, significant progress was also being made simultaneously on the origin of matter and the origin of the universe. Prior to the 20th century, astronomers had regarded the universe in general as a static phenomenon that had hardly changed at all. They assumed, therefore, that the material substance of the universe as a whole was constant, despite the fact that it was subject to movement and change. This was mainly expressed in the Steady State theory which was formulated around 1920 by the British physicist and astronomer Sir James Jeans (1877-1946), and which was later adapted in 1948 by other astronomers. According to this theory, the universe as we see it is eternal; it has always existed in this form and will always continue to exist in a similar way. This vision matched the materialistic world view perfectly; an eternal universe, consisting of an infinite amount of material particles that move about chaotically in the infinite empty space. "Only atoms and the void", as Democritus put it so succinctly.

The American astronomer Edwin Hubble (1889-1953) presented strong arguments against this theory. In 1927 Hubble made the important and remarkable discovery that the universe, according to his observations, was expanding outward. Hubble also discovered that this expansion occurred in all directions

The big bang, as we often imagine it. Of course, we cannot imagine what the big bang was like at all, but as it was some sort of explosion, such images are most common. In reality, the moment of the origin of the universe cannot be described at all and it cannot be simulated by any computer. This event exceeds our understanding and exceeds all human abilities. However, this also holds for the universe in its present condition. One thing is for sure, the big bang confirms the infinite mystery of our universe and it also confirms that the universe, by definition, does not have a material origin.

uniformly. This discovery had far-reaching consequences regarding the origin and structure of the universe in general. After all, if the universe is expanding then reasoning back in time; it must have sprung from one point. This conclusion led directly to the big bang theory.

The singularity and the big bang

The big bang theory is the theory about the origin of the universe that has gained the most support over the past decades. Cosmologists agree that the big bang took place about 13.7 billion

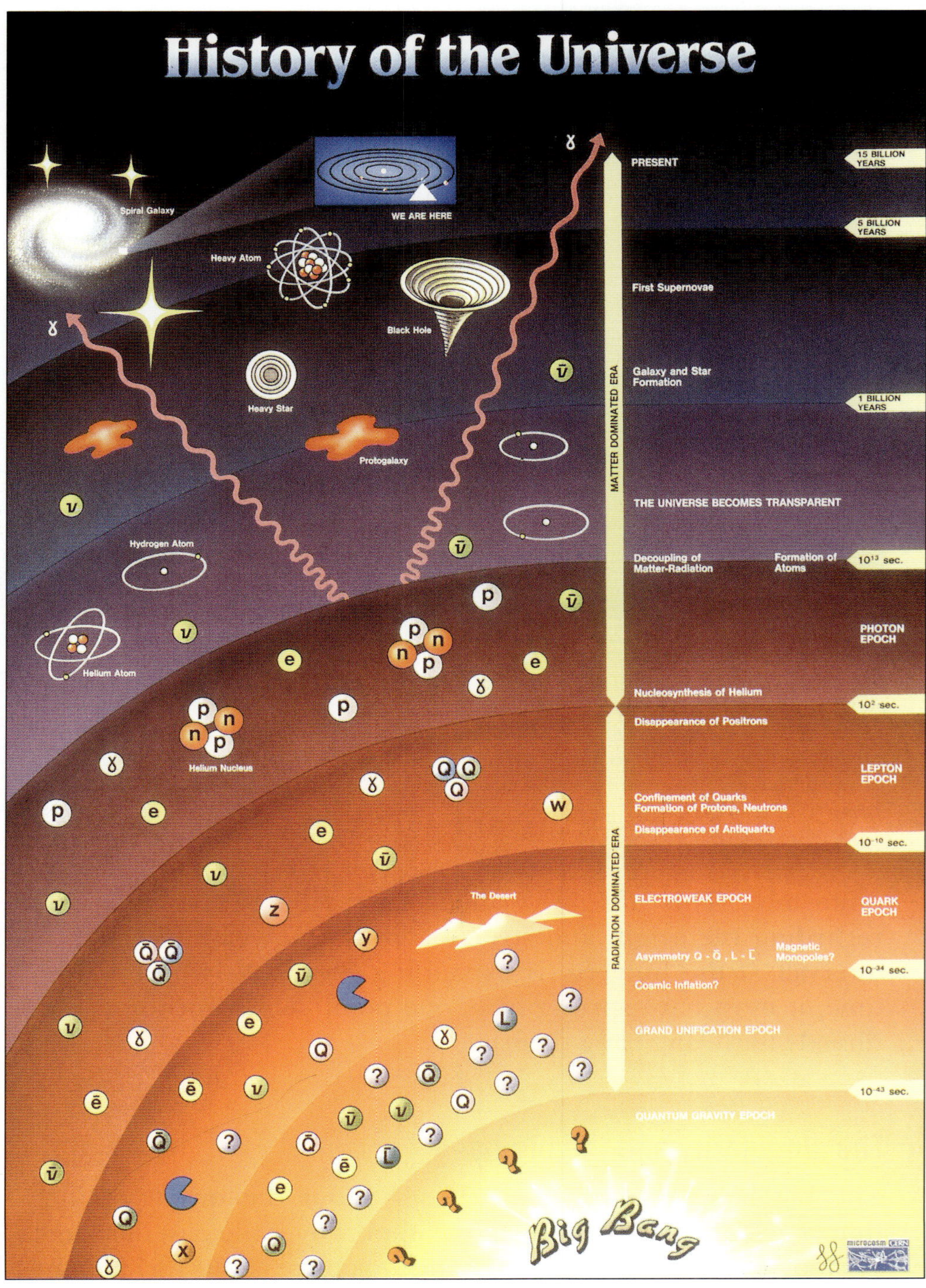
History of the Universe

Spiral Galaxy
WE ARE HERE
Heavy Atom
Black Hole
Heavy Star
Protogalaxy
Hydrogen Atom
Helium Atom
Helium Nucleus
The Desert
Big Bang

PRESENT
15 BILLION YEARS
5 BILLION YEARS
First Supernovae
Galaxy and Star Formation
1 BILLION YEARS
THE UNIVERSE BECOMES TRANSPARENT
Decoupling of Matter-Radiation
Formation of Atoms
MATTER DOMINATED ERA
PHOTON EPOCH
Nucleosynthesis of Helium
Disappearance of Positrons
LEPTON EPOCH
Confinement of Quarks
Formation of Protons, Neutrons
Disappearance of Antiquarks
ELECTROWEAK EPOCH
QUARK EPOCH
RADIATION DOMINATED ERA
Asymmetry Q - Q̄ , L - L̄
Magnetic Monopoles?
Cosmic Inflation?
GRAND UNIFICATION EPOCH
QUANTUM GRAVITY EPOCH
microcosm CERN

On the previous page is a diagram of the big bang, from the first 10- 43 second up to the first second, the third second, the first 380,000 years, and the first billion years up to now. In all these time units, unimaginable things happen, from the creation of gravity, radiation, the nuclear forces, the formation of quarks, the formation of protons and neutrons, the formation of photons, the dismantling of matter and radiation, the formation of atoms, the beginning of nuclear fusion reactions—where hydrogen is transformed into helium—up to the formation of stars, star systems, supernovas, and the universe as we know it today.

years ago. At that moment, existence and its related features such as time, space, matter and energy were created. Cosmologists furthermore agree that, regardless of the name 'big bang', this was not an actual explosion in the traditional sense of the word. The Big Bang was in fact a massive inflation that created an inconceivable amount of energy and mass at once. As we shall in later discussion, this inflation took place in a very controlled manner and it was absolutely not chaotic. The bestselling author Bill Bryson gives a good description of the big bang in his book *A Small History of Nearly Everything'*:

"In a single blinding pulse, a moment of glory much too swift and expansive for any form of words, the singularity assumes heavenly dimensions, space beyond conception. The first lively second (a second that many cosmologists will devote careers to shaving into ever-finer wafers) produces gravity and the other forces that govern physics. In less than a minute the universe is a million billion miles across and growing fast. There is a lot of heat now, 10 billion degrees of it, enough to begin the nuclear reactions that create the lighter elements – principally hydrogen and helium, with a dash (about one atom in a hundred million) of lithium. In three minutes 98% of all the matter there is or will ever be has been produced." [68]

The big bang itself arose from the singularity. Scientists describe the singularity as a condition of pure, undifferentiated existence. Energy, mass, time and space are infinitely concentrated at an infinitely high temperature in an infinitely small, timeless, space-less place. It is important to note at this point, however, that basically time, space, mass and energy do not actually exist in the singularity. This evokes two important questions; firstly, what exactly is the singularity, and secondly, does it have an origin itself or is it timeless and eternal? Science does not agree on the answers to these two questions.

The nature and origin of the singularity

With reference to the first question, we must conclude that there are multiple descriptions and definitions of the singularity. The description of the singularity as stated in the previous paragraphs is the one that is most commonly used: a condition of pure, undifferentiated existence, in which energy, mass, time, and space are infinitely concentrated at an infinitely high temperature in an infinitely small, timeless, and spaceless place. It is impossible to imagine this, but that is not so strange, for the singularity is a form of existence that basically exists outside of existence as we know it. We only have experience with the existence of things that exist, so anything that exists outside of existence is something that by definition we cannot comprehend or imagine. Bill Bryson has again given a good description of the singularity:

"It is natural but wrong to visualize the singularity as a kind of pregnant dot hanging in a dark, boundless void. But there is no space, no darkness. The singularity has no around around it. There is no space for it to occupy, no place for it to be. We can't even ask how long it has been there – whether it has just lately popped into being, like a good idea, or whether it has been there forever, quietly awaiting the right moment. Time doesn't exist. There is no past for it to emerge from. And so, from nothing, our universe begins. [69]

There is also a slightly different view on the singularity: a sea of undifferentiated and pure energy that also exists outside time and space. In this sea of pure energy, matter and antimatter are perfectly teamed up and they neutralise each other. Somehow this balance is disturbed, and therefore there is slightly more matter. From this imbalance the big bang arose, in which the surplus value of matter comes into being and is basically created. The difference with the standard big bang theory, which conveys the more common description of the singularity, is that such a sea of pure energy is not concentrated in an infinitely small point. This sea of pure energy is actually infinitely large. Even though this version is less usual than the infinite small singularity, it does, for example, agree with the ancient Indian Vedanta doctrine and the description of *Brahman*. According to the Vedanta doctrine *Brahman* is an undifferentiated sea of energy which is the foundation of all existence and which is the origin and cause of the universe. The Vedas describe *Brahman* as the infinite, eternal field of pure energy and pure consciousness which is the base of existence. Brahman is filled with all potential and actual properties and is the origin of the material universe. This description shows striking similarity with this version of the singularity.

Science is also divided regarding the answer to the second question, namely the origin and cause of the singularity. The famous scientist Stephen Hawking, and many others along with him, believe that the singularity arises from pure void and, therefore, from nothing. These scientists regard nothingness as the foundation of the singularity, and therefore of everything. Basically, this is an absurd and totally paradoxical thought that we will pay attention to in the later chapters of this book. At this point, we will just state that quite a few scientists say that 'the nothing' is the cause and starting point of the singularity. Others, however, regard the singularity as part of an eternal cycle. According to this theory, the big bang would be followed by a big crunch that is caused by gravity. The big bang makes the universe expand, but this expansion will gradually slow down because of gravity. At a certain point, gravity will pull the expanding universe back, causing it to implode into a new singularity. This singularity will again create a new big bang and this is again followed by a big crunch and this creates an endless cycle of creation and destruction. This is also called the bouncing universe, or the bang-crunch cycle.

The disadvantage of this theory is that it is not certain whether gravity is strong enough to slow down the expansion of the universe, or to cause an implosion. As we discussed previously, it appears from recent research that the expansion of the universe is actually accelerating. This would create yet another completely different scenario, namely the big rip. In this scenario all matter continues to expand in the infinite spatial void, in order to be eventually 'ripped apart' and accordingly almost completely dissolve. Even the atomic and sub-atomic structures will be ripped apart. The universe will freeze because the temperature will approach absolute zero. The reasoning for this is that the amount of matter is in the end always finite, while empty space is infinite. In this infinite void, matter will basically disappear and will become irrelevant. Basically, we can place the various scientific explanations regarding the origin of the singularity into three categories of possibility. The singularity is either part of an oscillating universe, or it arose from the absolute void, or it arose from an undifferentiated, ultimate form of existence - a sort of existence that exists beyond the boundaries of existence, without any features such as energy, mass, temperature, time or space.

All of these explanations and their variants assume a condition before the singularity, in which the origin of the singularity is non-causal. This also holds for the transformation of the singularity into the big bang; this event is also non-causal, and it could not have been any other way. Time, space, matter, energy and the laws of nature after all do not exist in the singularity and the form of existence prior to the singularity. Without laws there cannot be a causal connection between one event (the condition of the singularity) and the following event (the big bang). The astronomer Fred Hoyle, who has been cited before, describes the big bang as follows:

"All that we see in the universe of observations and fact, as opposed to the mental state of scenario and supposition, remains unexplained. And even in its supposedly first second

> *the universe itself is acausal. That is to say, the universe has to know in advance what it is going to be before it knows how to start itself. For in accordance with the big bang theory, for instance, at a time of 10 $^{-43}$ seconds the universe has to know how many types of neutrino there are going to be at a time of one second. This is so in order that it starts off expanding at the right rate to fit the eventual number of neutrino types."* [70]

This also implies that the Big Bang must have been a controlled and organized process of unbelievable complexity. From the first measurable fraction of a fraction of a billionth of a second of the Big Bang, at every subsequent step of the process very precise events had to take place. The inflation theory was first propounded by Alan Guth, in which he proposed that the massive inflation started at 10 $^{-43}$ seconds after the creation, doubling in size every 10 $^{-34}$ seconds. At each of these tiny slices of time intervals another feature of the universe came into being, first gravity, then electromagnetism, and then the strong and weak nuclear forces. After 10 $^{-30}$ seconds the universe expanded, according to the theory, from nearly nothing to a 100 billion light years in size, filled with sub-atomic particles, the stuff that eventually made up atoms and matter. Both the rate of expansion as well as the values of the various just created forces are precisely tuned for a successful outcome, i.e. a functioning universe.

The famous scientist and Nobel Prize winner Steven Weinberg said:

> *"The standard model sketched above (the big bang) is not the most satisfying theory imaginable about the origin of the universe. (...) There is an embarrassing vagueness about the very beginning, the first one-hundredth of a second or so. Also, there is the unwelcome necessity of fixing initial conditions, especially the initial thousand-million-to-one ratio of photons to nuclear particles. We would prefer a greater sense of logical inevitability in the theory"* [71]

Steven Weinberg

Furthermore, one can wonder in all cases why reality would want to create itself, especially if the only reason for existence would be existence itself. From the perspective of a mechanistic world view, existence is far less efficient than non-existence. This matter of efficiency will be thoroughly discussed in later chapters. Nevertheless, the big bang is an absolute fact and the best possible explanation regarding the origin of the universe; despite the fact that the scientific explanation regarding the origin of the big bang and the singularity is not satisfactory and despite the fact that it is surrounded with vagueness. This vagueness is the logical consequence of the complexity and the incomprehensibility of the singularity and the big bang, but it will become far less vague when it is placed within the scope of a non-mechanistic and non-materialistic worldview.

God and the singularity are very much alike

We actually see—and we have just concluded this—that the advanced scientific insights of the 20th and 21st century implicitly refer to the existence of God. The singularity is a very good example of this. The properties that scientists attribute to the singularity do not really differ from the key properties that are traditionally attributed to God. The singularity is described as an 'infinitely small point, beyond space and time, with an infinite amount of energy that exists with an infinite density and an infinitely high temperature." In this description reference is made several times to infinity and the factual indescribability of this condition of being. How can an infinitely small point contain an infinite amount of energy and mass? The infinitely large is squeezed into the infinitely small here. And how can an infinitely large amount of energy that is also squeezed into an infinitely small point have an infinitely high temperature? Temperature is determined by the movement of particles in relation to each other. But the singularity, as its name suggests, does not have any parts. Being one system, it cannot have 'temperature'. What this is really about is the fact that the singularity contains an infinite amount of energy, for energy causes movement and therefore temperature. Temperature, however, is of crucial importance, for in all explanations about the singularity and the big bang it appears that the energy in the system—that is, the temperature— at a certain point cannot be restrained or controlled anymore. At that moment, a massive explosion, or rather inflation, takes place and the big bang comes into being. Most explanations at least hint at such a series of events. Of course, there is no causal connection between the singularity and the big bang that follows it, for within the singularity the laws of nature do not yet exist. Therefore, the big bang is a non-causal event. On the other hand—and here things become contradictory—the infinitely high temperature is implicitly mentioned as the reason why the singularity stops being a singularity, transforming itself into the universe through an unimaginably powerful and violent process. In this explanation, temperature is regarded as a kind of cause of the big bang. Besides extremely high temperatures, other reasons are also provided to explain the miraculous transformation from singularity into a universe. As discussed previously, some theories suggest that there was a small imbalance between the matter and antimatter contained within the singularity, which imbalance led to the big bang.

These descriptions are, of course, unimaginable for humans. The words that are used here do give us an idea of what is meant, but they become incomprehensible when you think more deeply about the logical consequences. The definition that scientists give of the singularity strongly resembles the definitions and description of God in some spiritual and religious traditions. God is described as a creature that is beyond human comprehension, infinite, eternal and indescribable. In the Upanishads, which are part of the Vedic literature in India, God is called *Brahman* and is described as *Acintya*, or, in other words, unimaginable. The Vedas describe *Brahman* as the infinite, eternal field of pure energy and pure consciousness which is the base of existence. *Brahman* is filled with all potential and actual properties and is the origin of the material universe. It is furthermore described as 'the biggest of the biggest and the smallest of the smallest" [67] and as "he who moves and yet does not move"[68] and as "he who is close and near, imminent and transcendent" [69]. These are descriptions that, metaphysically and also mysteriously, are no less than the descriptions of the singularity.

In many ways the singularity of the big bang cosmology resembles the definition of God. However, despite these similarities, there are three important differences:

1. **God is explicitly eternal, while the singularity is only implicitly eternal;**

2. **God is infinitely great and infinitely small, while the singularity is only infinitely small;**

3. **The singularity is dumb and unintelligent, while God is infinitely intelligent.**

These differences, however, are mainly of a semantic nature.

The difference between infinitely small and infinitely large becomes semantic when the infinitely small also carries in itself the infinitely large, then basically the infinitely small becomes infinitely large. The difference between the explicit and implicit eternity is also partly of a semantic nature. The singularity is described as that which is nevertheless the creator of time, but simultaneously itself has a sort of beginning. This 'sort of beginning' is the nothing that transforms itself into the singularity. In this way, 'nothing' is elevated into the eternal, timeless source from which the singularity and the universe have arisen. Since 'nothing' does not exist and therefore cannot produce anything, the singularity itself becomes implicitly timeless. The real and most important difference between the singularity and God, therefore, is that the singularity is ignorant, while God is intelligent. If we were to attribute intelligence and consciousness to the singularity, then the singularity and God would no longer differ from each other.

The implications of the big bang

The big bang theory also gives us an explanation about the further development of the universe and the evolution of matter in the universe. Matter in the universe goes through an evolutionary process during the expansion following the big bang in which sub-atomic particles, atoms, an entire series of chemical elements and stars, solar systems, star systems, clusters and super clusters are created. According to scientists the first three minutes, or even the first seconds after the big bang were of crucial importance to shape the universe in the exactly right manner. In summary, the big bang theory has four important consequences and all four of them undermine the materialistic world view:

- First of all, the big bang theory confirms that the universe has a cause. This contradicts the former mentioned Steady State theory that assumes that the universe in its present shape is eternal. Furthermore, this cause must have existed outside the universe itself.

- Secondly, because of the limited life span of the universe, the teleological argument, even though it is not relying on this, resurges with additional strength. This has already been substantiated in the previous chapter.
- Thirdly, the big bang theory confirms that the origin of matter itself is non-material. The singularity and the form of existence from which the singularity comes, are non-material forms of existence by definition. I have already referred to the definitions of the singularity and also to the possible explanations of the origin of the singularity. These explanations vary from 'nothing' to a pure, undefinable form of existence. None of these descriptions look like the world of hard, tangible, perceivable matter, which the empiricists—the founders of the scientific method—had in mind. It is clear, therefore, that the cause of matter, stands outside and above matter itself. Moreover, the definitions of God and the singularity can hardly be distinguished from each other, except for the attributes of intelligence and consciousness.
- Lastly, the big bang theory states that the universe came into being from something that is a fundamental unity. The singularity—and the term already suggests this—is an absolute unity. This fact fundamentally contradicts the materialistic worldview, in which pluralism dominates and in which the unity of pure being is ripped apart by the emergence and insertion of empty space and the void. Here we observe a peculiar reversal, whereby pure being produces non-being, instead of non-being producing being. The latter notion of course being seen as the viable explanation for the origin and source of the singularity.

The big bang theory therefore has far-reaching consequences as far as our current understanding of the universe is concerned. Above all, it undermines the purely materialistic view of the universe.

Materialism is incomplete

In conclusion, it appears from the aforementioned discussion that materialism is inadequate, incomplete, and fundamentally incorrect. Of course, matter exists and the various mechanical systems that are the result of the interaction between these forces and the material particles also exist. The New Physics, for example, is not a denial of Newton's laws, but rather an important addition to it. In other words, the laws of Newton were not incorrect, but they were incomplete. They become incorrect if it is argued that the Newtonian worldview represents the total explanation of reality, instead of just a partial one. This also holds for the materialistic worldview. Materialism and its mechanistic foundation is correct in certain sub-sections, but it is also incomplete. When philosophers and scientists claim that this describes the whole of reality, then materialism is incorrect. However, if materialism does not adequately describe reality, then what should replace it? One of the key insights provided by the New physics is that the universe consists of two forms of being, which are mass and matter on the one hand, and energy on the other hand. While matter represents the world of the familiar, being tangible, localised, containing mass and governed by mechanistic laws, it appears that the world of energy is completely different. Energy is, in many ways, the opposite of matter, despite the fact that matter is a transformation of energy. Energy is not directly tangible, it is not localised, does not have mass, and is not mechanistic. Above all, energy appears to be capable of creating very complex and intelligent structures, and it has properties that can be described as profoundly mysterious. Furthermore, according to the teleological arguments presented in our earlier chapters, this level of complexity and organization cannot have happened by chance which implies that energy is intelligent. The world of matter is therefore part of a much larger energetic reality that is totally different from what we suspected initially.

Based on all the arguments and scientific developments presented so far, we cannot avoid the conclusion that the universe is much more complex and mysterious than we could have ever imagined. Our understanding of the universe is no longer a simple, scientific matter that is based on direct, hard, perceivable data and experiments. The relative simplicity of materialism gives way under the weight of so much data and information that contradicts it. Therefore, we cannot escape deeper philosophical considerations and the necessity for metaphysical and logical analyses. In some respects modern scientific theories look more like the metaphysical speculations of the philosophers of the Enlightenment. Whether describing relativity, quantum theory, string theory, big bang cosmology, cosmic inflation or the singularity, none of these theories, no matter how elegant, intellectually brilliant, or logically cohesive they may be, are actually based on directly observable realities.

Of course we can distinguish between various degrees of speculation. Direct observations or inferred observations that occur in the present (for example, in a particle accelerator) have a stronger scientific impact than speculations about the origin of the universe 13.7 billion years ago. While theories concerning the origin of the universe may have been developed based on observations in the present, such as Hubble's discoveries, it is still a massive time jump from the present to 13.7 billion years ago. A lot could and will have happened in between that we do not know anything about. The history of science has clearly taught us that scientific theories have to be constantly adjusted and sometimes even have to be completely replaced by new and better theories. Technological progress and the continuous improvement of our measurement and observation equipment are key reasons. Other reasons are new generations of scientists and the brilliant new insights they produce. Albert Einstein is a perfect example of this. He developed his theory of relativity mainly through thought experiments and mathematical analyses. Only years later were his propositions confirmed by scientific experiments and observations.

Even though science has gathered an unimaginable amount of detailed data, it is also the case that a lot of this data is extremely complex and even incomprehensible so that it does not necessarily lead to knowledge and understanding. Information and knowledge are two separate issues, whereby an increase in information does not necessarily mean an increase in knowledge. In fact from a philosophical point of view it often turns out to be the opposite, since the overwhelming amount of very detailed complex and highly technical information can cause one to lose sight of the bigger picture. As a result the increase of information leads to an increase in scepticism and a resignation to the notion that an ultimate understanding of the universe is unobtainable. That is incorrect though, for fundamental scepticism and nihilism are unwarranted. Reality exists, we are aware of that reality and since our consciousness is part of that reality, it can be comprehended. Our disadvantage is that we are extremely small and insignificant, whereas reality itself is infinite. Yet, even though reality is infinite it has a holographic structure, whereby each part of reality carries within it the entirety of the whole of reality. Furthermore, each part of the whole is qualitatively one with the whole and therefore knowledge of the part can provide at least qualitative knowledge of the whole, and give as an understanding of the big picture.

"With respect to this, 'knowledge' can be compared to a map; a map with which we navigate our journey through an unfamiliar landscape. The map differs quantitatively from the actual landscape. The map is only 80 by 80 centimetres and the landscape itself is tens or even hundreds or thousands of kilometres in size. However, a good map will give the traveller a qualitatively correct image of the actual landscape —the big picture- and will therefore navigate the traveller in a correct way through the landscape. Knowledge works in a similar manner, but only if it is proper and correct knowledge."

The correct map, the correct knowledge can help the traveller navigate through his life journey and through this complex universe to the correct and desired destination. What this destination is, how we get there and which knowledge is needed to get there is the topic of this book. To acquire this knowledge we use the most advanced scientific insights from particle physics to cosmology. But we will also take advantage of the brilliant philosophies from ancient India such as the 6000 years old Vedanta doctrine. Vedanta doctrine formed the basis of two world religions, Hinduism and Buddhism and it is a doctrine that has also impressed many western philosophers and intellectuals because of its refined ideas and intellectual depth. This undertaking may sound ambitious, but the search for comprehensive knowledge is ambitious in itself. It is, after all, the ultimate goal of science too. The search for the 'unified field', the fusion of relativity with quantum physics, understanding the origin of the universe and the building blocks of matter, is no less than the search for an comprehensive understanding of reality and how that relates to our own existence.

In the next chapters the third and fourth propositions of atheism will be discussed, in the chapters after that we will be focussing on the larger picture, and the implications of a universe, both from a scientific and a human perspective, that is created, controlled and guided by intelligence.

Chapter 6

THE ORIGIN OF EXISTENCE HAS TO BE 'SIMPLE'

"Everything should be made as simple as possible, but not simpler." [72]

Albert Einstein

The third atheistic proposition – the cause of everything must be simple

The third proposition of atheism is based on the argument that the ultimate cause of the universe cannot be complex: it must be simple, and not just simple, it must be ultimate simple. In this chapter, we will examine this argument and show, firstly, how this argument is not based on any type of scientific observation or method, it is just an assumption. Secondly, by using the laws of logic and causality, we will show that it is an illogical and irrational assumption that runs contrary to the scientific method and pursuit of knowledge as we know it. We conclude by demonstrating that our complex universe must have a cause that is at least equally complex, i.e. that it contains all the energy, substance, properties, intelligence and organized complexity both in an actual and potential manner.

In his book The God Delusion Dawkins regularly refers to the fact that if God existed as the intelligent creator of a complex universe, then God himself would have to be very complex and thus he would have to have been created himself.

"I have alluded to it several times already. The whole argument turns on the familiar question 'Who made God?', which most thinking people discover for themselves. A designer God cannot be used to explain organised complexity, because any God—capable of designing anything would have to be complex enough to demand the same kind of explanation in his own right. God presents an infinite regress from which he cannot help us to escape." [73]

A chapter later Dawkins repeats this argument in a report about a conversation that he had with some theologians:

"Time and time again my theologian friends came back to the point that there has to be a reason why there is something instead of nothing. There has to have been a First Cause of everything and that we might as well give it the name God. Yes, I said, but it must have been simple and therefore—whatever else we call it —God is not an appropriate name (unless we very explicitly divest it of all the baggage that the term 'God' carries in the minds of most religious believers). The first cause that we seek must have been the simple basis for a self-bootstrapping crane which eventually raised the world as we know it into its present complex existence." [74]

In the above quotes, Dawkins asserts that the universe has a cause, and indeed, the scientific insights of the past decades confirm this assertion. However, Dawkins also asserts that this cause must be the simplest possible cause, devoid of any complexity. He makes this claim with a great deal of certainty but without giving any further explanation as to why this should be the case. Factually, this proposition is nothing more than an assumption based on a personal preference.

Indeed, the necessity of a simple cause of all causes is based on an intuitive assumption that partially reflects some of the methods used in science, a method called reductionism. Reductionism, as the term implies, is a method of reducing and dissecting complex structures into its simpler components, all the way to its fundamental building blocks. In essence there is nothing wrong with this method, but as will be explained later in this chapter, reductionism cannot be applied to all situations. Reductionism becomes a problem if the sum of all parts becomes greater and contains more properties than the parts themselves.

Furthermore, one can wonder what this 'simple' thing should be. As we have concluded previously, the singularity has been suggested to be this 'simple' cause, but effectively the

Sir James Jeans (1877 – 1946)

singularity does not qualify given the definitions of singularity as was discussed in the previous chapter: 'an infinitely small point beyond time and space in which an infinite amount of mass and energy is compressed at infinite density at an infinitely high temperature.' This description not only implies something unimaginable of mystic proportions; it also indicates that the singularity is a reality that is infinitely complex and in which all existing things as we now know them in the universe, such as time, space, mass, energy, and all forces such as gravity, electromagnetism, and the nuclear forces are present in a latent form. This certainly does not qualify as something of ultimate simplicity. In fact the only 'thing' that is ultimately simple is nothingness, nothing is simpler that nothingness. The problem is that nothingness does not exist, which makes nothingness as the ultimate cause of all causes extremely problematic. We will discuss this problem in detail in chapter 8. At this stage, however, we will review the current scientific evidence and theories about the cause or origin of the universe.

The origin of the universe

Presently the prevailing scientific view is that the universe has a beginning, in the form of the big bang, preceded by the singularity. But this has not always been the prevailing view. Up to the beginning of the 20th century, astronomers and physicists, Einstein among them, regarded the universe as a static, constant phenomenon, everlasting and without an explicit cause. This was called the Steady State theory, to which we referred to in the previous chapter. This theory was formulated in about 1920 by the astronomer Sir James Jeans. This theory was soon to be challenged by a number of cosmologist, in particular by Hubble, who had discovered in 1927 through observations that the universe was in fact expanding. Hubble published his findings in 1929, and they provided the empirical evidence that eventually led to the formulation of the big bang theory. Supporters of the Steady State theory, such as Fred Hoyle, Thomas Gold and Hermann Bondi did not admit defeat so easily, and they adapted and modified the Steady State theory by acknowledging that the universe was indeed expanding. To compensate for the expansion they introduced a new theory called the 'perfect cosmological principle' which entailed the notion that the speed of the expansion was in accordance with and proportionate to the continuous creation of new matter. In this way they were still able to preserve the notion of a stable and static universe.

Following new observations in the 1960's, however, this adapted theory of the Steady State theory was also refuted. These new observations related to the discovery of so-called background radiation. Supporters of the big bang theory had predicted the existence of this background radiation as a remnant and an echo of the big bang. Background radiation was discovered purely by chance by two radio astronomers, Arno Penzias and Robert Wilson in 1965, who went on to receive the Nobel Prize in 1978 for their discovery. The background radiation perfectly met the expectations and nature of this echo. This discovery became the final blow to the Steady State theory, and the big bang theory was subsequently recognized as the predominant explanation with respect to the cause of the universe. One of the important implications of the big bang theory is the fact that the universe has a beginning and that it originated from 'something else'.

The formulation of the Big Bang theory including the notion of an expanding universe, has been one of the most spectacular scientific discoveries of the 20th century. The concept that the universe had a beginning had so far mostly been associated with religious belief and in the Christian world with the Biblical story of creation in the Book of Genesis. Now the creation story was shifted to the domain of science, a truly revolutionary development. Usually things that have a beginning, also have an ending, and while there is common agreement today about the origin of the universe, scientist are divided over its final destination. Will the universe continue to expand forever into the infinite and eternal void, or will the expansion come to a halt and either stabilize or collapse in on itself. Initially, the 'big crunch' scenario was suggested whereby the expansion of the universe would in time come to a standstill because of the effect of gravity. After that, again because of gravity, the universe would start contracting, resulting in an implosion. In the end, the universe would return to its original position: the singularity, which in turn might lead to a new big bang, the so-called bouncing universe. Recent observations have confirmed the opposite, however, namely that the expansion of the universe is accelerating. Therefore, in time all matter, up to its atomic structures, will be ripped apart: the so-called 'big rip' scenario. In the end, according to this scenario, the universe would undergo a freeze death. While the future of the universe, based on current scientific insights is unclear, scientists have a clear view on its past and its origin. The Steady State theory is out, and the Big Bang is in.

The Steady State theory was unlikely to succeed for a number of other reasons. The most important reason is in fact Einstein's theory of general relativity. According to this theory gravity acts on all matter in the universe as a force of attraction and only in one direction. As a result all the matter in the universe would eventually contract and implode, and therefore the universe could not possibly be stable. Einstein recognized this problem shortly after publishing his theory of general relativity, even though he was at that time an adherent of the Steady State theory. He realised that his own theory would undermine the notion of an eternal and stable universe, and in 1917 he came up with a solution that became known as the "Cosmological Constant". The Cosmological Constant is in some respects similar to the "Perfect Cosmological Principle" referred to above, in that it assumes the creation of new matter in this instance in order to counteract the inward pull of gravity. In this way the universe could still be stable and eternal. Some years after Hubble's discovery of the expanding universe in 1927, Einstein had a meeting with Hubble in Pasadena California in 1931. There they discussed the notion of the expanding universe, and Einstein now became convinced that the universe was not static. Around the time of Hubble's discovery, also in 1927, the Belgian priest Lemaitre had published his theory of the primeval atom, in which he suggested that the entire universe had sprung from a single, infinitesimal primordial "atom" that expanded into the universe. Interestingly Lemaitre had succeeded on the bases of mathematical calculations to formulate the expansion of the universe and even the rate of expansion that became later known as Hubble's constant. He did this prior to the publication of Hubble's findings, and according to some historians Lemaitre did not receive enough credit for his role in the formulation of the Big Bang theory. His primeval atom theory, supported by Hubble's observations eventually evolved into the Big Bang theory. As stated, Einstein accepted this model of an expanding universe with an origin comparable to Lemaitre's "primeval atom", and he abandoned the Steady State theory. Later he would call his notion of the Cosmological Constant his "biggest blunder".

There are other reasons why the Steady State theory was unlikely. The universe and everything in it consists of parts that are subject to perpetual change, both internally as well as in their interactions with each other. Fundamentally, therefore, the universe is unstable, both at a micro level as well as at a macro level. For instance at the micro level it was assumed until the end of the 19th century that atoms were indivisible and unchangeable, and therefore stable. This assumption turned out to be incorrect; atoms were divisible and could be split up into even smaller sub-atomic particles. The search for the ultimate building block of matter, the subject of particle physics has led to the discovery of some 61 sub and sub-sub-atomic particles. Of these particles quarks come close to having the status of being the ultimate particle, although some would argue that at that level the notion of 'particle' no longer fits the bill. Research into the very complex world of the sub-atomic micro cosmos has led to the development of quantum physics, a theory that has been discussed in more detail in the previous chapters. One of the most important insights of quantum physics is that the difference between particles and non-particles fades. Particles sometimes appear to be energy fields whilst at other times they behave like particles of matter. Furthermore, it appears that this interaction between particles and fields is not predictable. It is clear that the sub-atomic world is not simple, nor is it stable.

At a macro level, cosmologists have discovered that the universe, with its solar systems, star systems, clusters and super clusters, is an extremely changeable, unstable, and dynamic environment. Solar systems like ours are transitory. Cosmologists have calculated that our sun, now 5 billion years old, will have burned up her nuclear fuel in another 5 billion years and will therefore die after 10 billion years of loyal service. In the first phase of this process of dying, the sun will become hotter and hotter. The sun will physically expand, and the inner planets of our solar system, including earth, will be completely engulfed and incinerated. In the second phase of her process of dying, the sun will implode and end up as a white dwarf: a lifeless, ice-cold lump of matter aimlessly floating in space. With this, the fate of the sun and our solar system is sealed forever and they will cease to exist. From atoms, up to solar systems, up to star systems, up to the entire universe itself, everything is unstable and subject to constant change. Within this fundamentally unstable structure of the universe, the Steady State theory is not a very plausible explanation.

The material cause and the efficient cause

The big bang theory has established the fact that the universe does not exist eternally. The universe has a cause, and it has sprung from something else, the singularity. In the chain of cause and effect it is important that we understand the underlying laws that rule this chain. The laws of cause and effect are very important, for they govern the universe, they govern science, they govern the rules of logic and they govern our lives. In Western philosophy, Aristotle was one of the first philosophers to describe and analyse the laws of cause and effect. He observed that certain events resulted in other, predictable events. He analysed four kinds of causes for each effect:

1. *the material cause,*
2. *the formal cause,*
3. *the efficient cause,*
4. *the final cause.*

When a potter, for example, makes a pot out of clay, then the clay is the material cause of the pot, the vision and craftsmanship of the potter is the formal cause, the actions performed by the potter the efficient cause, and the purpose for which the pot was made is the final cause. Later philosophers such as Francis Bacon reduced the categories of causes into only two: the material and the efficient cause. In his view the formal, efficient and final cause could be combined into one, the efficient cause. This is logical, since both the formal cause as well as the final cause spring from the consciousness of the potter; namely the vision of the yet-to-be created pot in the mind of the potter, as well as the purpose for which the pot was designed and made. This also holds for the efficient cause. This cause was also generated by the consciousness of the potter where the willpower of the potter made him decide to take action and to manufacture a pot.

Regarding natural phenomena the matter is more complicated, since the efficient cause is mostly an impersonal force, such as the laws of motion, which bring about change or movement. Regarding chemical reactions, for example, putting

Aristotle

together hydrogen atoms and oxygen atoms under certain circumstances leads to the production of water molecules. The material cause of the water molecules are therefore the hydrogen and oxygen atoms; the efficient cause is the process of change that makes these atoms join together correctly. When extending the principles of cause and effect to the origin of the universe itself, we have to go back to the Big Bang, and the singularity. At the time of the big bang the entire universe including energy, time, space and matter came into being, and from then on everything was put into motion. The efficient cause is the energy and the forces that were released by the big bang that made the sub-atomic particles move and cause them to interact with each other. The material cause is the singularity from which matter and energy came into being that make up the stuff that constitutes the universe. The singularity is therefore both the material and efficient cause of the universe, which immediately brings about the question what has caused the singularity, and can the singularity qualify as the eternal cause of all causes. This question has already been discussed in chapter 5 where the singularity has been compared to God, and where it became clear that the singularity falls short in significant ways. At this point it is important to understand how the laws of cause and effect function and what we can conclude from them.

The principle of causality: the effect cannot be greater than the cause

The idea that matter can organize itself, or that the universe has appeared out of nothing by means of the big bang and singularity contains a fundamental contradiction. This contradiction is included in the basic principle of all scientific and logical arguments, namely the **principle of causality**. In Indian philosophy the principle of causality, or cause and effect, has been thoroughly discussed and examined. In Vedanta doctrine, this principle is called '*parinama vada*' in Sanskrit. This is a fundamental principle, which states that the effect can never be greater than its cause. *Parinama vada* means that each effect is actually or potentially contained within its cause. The principle of causality can therefore be defined as follows:

> *According to the principle of causality, an effect, whether it is a material effect or an efficient effect, can never be greater than its material or efficient cause. It is not possible to create something out of nothing; something cannot come into being out of nothing.*

Our common sense as well dictates that something cannot come into being out of nothing. After all, nothing does not exist, so how can it possibly produce anything? The Roman philosopher Lucretius (99-55 BC) stated:

"Nothing comes from nothing." [75]

In chapter 8 and 9 the relationship between being and non-being will be discussed in much more depth.

The principle of causality forms the bases of the scientific method and also forms the bases of the laws of logic. These laws were also formulated for the first time in Western philosophy by Aristotle and were later used by many great thinkers who adapted certain aspects. The laws of logic consist of three basic rules:

1. **The law of identity: A is A; in other words, something that is, is.**
2. **The law of the non-contradiction: A is not B; in other words, A is not not A; or in other words, something that is not, is not.**
3. **The law of the excluded middle: either A exists or A does not exist; in other words, A has to exist or it does not exist. This law states that something must either exist or not exist and that there is nothing in-between existence and non-existence.**

These laws seem so obvious and elementary that it is strange to formulate them like this and to actually write them down. The reason they seem so obvious is —and this is crucial— that they are based on fundamental observation, i.e. observation that is general and universal in the broadest sense of the word. We perceive that things are just as they are, we perceive that things are different from each other and we perceive that things either exist or that they do not exist. This, in all simplicity, sums up the three laws of logic. Nevertheless, in practice it appears that the greatest misconceptions in science and philosophy coincide with an incorrect application and an incorrect interpretation of these laws. The laws of logic seem to be stating the obvious, but on second thought this may be not as obvious as you think.

The laws of logic mirror the law of causality, since both laws basically claim the same thing. The core of these laws is the rigid distinction between existence and non-existence and the fact that existence cannot arise from non-existence. Denying the principle of causality means we accept irrationality as a foundation of knowledge. If we were to do that, we would be throwing away all scientific research of the past 400 years and all philosophical knowledge of the past 5000 years. The entire system of scientific knowledge and human logic is based on the relationship between cause and effect. Eliminating a meaningful relationship between cause and effect by reducing the cause to nothingness makes practicing science and logical thinking completely impossible.

From a scientific perspective the law of causality is formulated in the law of conservation of mass and energy, expressed in the world-famous formula $E=mc2$. This law gives us a good insight into the fundamental nature of existence. In simple words this law states that mass can be transformed into energy and vice versa, but they can never be destroyed. Therefore, they are both eternal and timeless. This law of nature also determines that something cannot come from nothing. Mass can be transformed into energy and energy can be transformed into mass, but transformation is different from creating something out of nothing. Even if the amount of mass and energy is infinite, then this still does not change the validity of this law: mass or energy cannot be created out of nothing.

Indeed, this fact had already been determined before Einstein, by James Joule, who formulated the laws of thermodynamics in 1849. These laws formulate the link between heat-energy and mechanical, electrical, and chemical energy within a closed system. The first law of thermodynamics, the law of conservation of energy, is of particular importance here. This law determines the amount of energy in an object as a consequence of its movement in relation to other objects. This energy can manifest itself as heat, light, or physical motion. According to the first law of thermodynamics, matter can be transformed into energy, for example, in the form of a chemical reaction. This law also applies to the transformation of one type of energy — for example heat— into another sort of energy, such as physical motion. In all these transformations, however, no mass or energy is created or destroyed. $E = mc^2$ goes even further, for it does not limit itself to the motion of one object in relation to another object. Einstein's law refers to all the energy contained in an object, including its mass, even if this mass is not in motion. Furthermore, the conservation of mass and energy must also be applicable to the properties of mass and energy, such as odours, temperature, colours, tastes, and shapes, for these properties are never observed separately from their underlying substances and are an integral part of them. This also means, therefore, that the properties of mass cannot arise from nothing.

This principle is applicable even if we assume that properties are actually the consequence of the interaction between consciousness and the observer of these properties and its underlying substance. Many scientists and philosophers believe that only the substance of an object is objectively real, while properties such as colour, sound, taste, and smell are basically the result of the interaction with a conscious observer. Philosophically speaking, this is a form of idealism, which we discussed in previous chapters, which means that external reality, or part of it, is a product of the consciousness of the observer. This form of idealism, in which objects are divided into substance and properties, has become the predominant philosophical point of view of classical physics. It means that the universe is divided into two parts; on one hand, an objective reality that consists of substance, form and motion, and on the other hand, a reality that is subjective, consisting of properties such as colour, sight, sound, taste, smell, warmth, cold, softness, etc. According to this theory, the subjective properties are indeed caused by the objective part of reality, such as mass, shape and motion, but they do not have an objective existence independent of the conscious observer.

In quantum physics a lot of attention is being paid to the relationship between consciousness and the observation of external objects. Some prominent scientists suggest that not only the properties of an object are the result of a conscious observation but also that the substance, such as the mass of an object, is the result of the interaction with a conscious observer. In a certain way this is correct; the observer and the object of observation are connected to each other in a special way. It is also correct that conscious energy is the foundation of reality and that matter is a transformation of this energy. On the other hand this does not imply that the individual consciousness of an arbitrary observer is the creator of all that he observes. That conclusion would be incorrect and obviously at loggerheads with our day to day experiences. The objects around us, such as the colour of the wallpaper, present themselves to an observer without the observer having the ability to influence it. If the observer does not like the colour of the wallpaper then he/she will have to do more than just wish it were different. The observer is not able to change the colour of the wallpaper just through the interaction with his/her consciousness.

The relation between substance and properties and the relationship between the observer and object of observation is a very interesting and important discussion indeed. Unfortunately, it is also a discussion that is outside the scope of this book. The key point that we are trying to convey at this stage is that both substance and properties cannot arise out of nothing.

Simple cause is just an assumption

In the end, Dawkins' proposition that God, as a complex being, must have been created himself is circular reasoning. Basically, Dawkins is convinced of the fact that the origin of the universe is ultimately simple and that it can only be simple. This conviction, however, is based on an assumption and nothing more than that. And from this assumption, Dawkins concludes that a complex being, such as God, cannot be the cause of the universe and therefore God cannot and does not exist.

There is, however, no compelling or logical reason for this assumption. In the end, existence at its most fundamental level, whether it be energy or matter or both, exists eternally and timelessly and eternal existence per definition has no cause. In the chain of existence there must be a cause that does not have a cause itself. While $E=mc^2$ shows that energy can be converted to mass and vice versa, neither can be destroyed. That implies that something does not and cannot arise from nothing, and that the foundation of existence—energy—is eternal and does not have a cause. That which has no cause can be basically anything, exactly because it does not have a cause. The lack of a cause makes the ultimate existence completely free of any compelling reason for a specific kind of existence. This implies that the ultimate existence could be very simple, but it could also be infinitely complex, or, as is the case could be both simple and complex simultaneously. We can only determine the nature of existence and the nature of its cause by observing existence itself and by observing its properties. At this point our observations are evident; existence is extremely complex, and this complexity contains a high degree of organization. It is then logical to conclude that the cause of this complex reality must contain all of the elements of organized complexity itself.

Dawkins, however, is an adherent of the evolution principle, and not just the Darwinian form of biological evolution. In chapter 4, the theory of evolution of Darwin was discussed extensively, and when we say, 'the evolution principle' we mean evolution in the broadest sense of the word. Evolution literally means 'gradual development', whereby an object develops properties in the course of time that it did not possess at the beginning of its development. The evolution principle implies that during the process of gradual development, new properties emerge essentially out of nothing. The evolution principle is also applied to the origin of the universe and the development of the universe since its origin. In its most extreme interpretation, some scientists have proposed that existence itself, the singularity and the subsequent universe, literally came into being out of nothing.

To quote Stephen Hawking:

"Because there is a law such as gravity, the universe can and will create itself from nothing"

In a slightly less extreme interpretation of the evolution principle, it is proposed that an eternally existing simple substance in due course of time develops into something more complex with entirely new properties. In this scenario not the underlying substance but only its properties come into being out of nothing, which is in fact one of the essences of the Steady State theory. The emergence of atomic structures and chemical elements from primordial material particles are often cited by scientists as good examples of the emergence of new properties out of nothing. The same applies to the emergence of life and consciousness from essentially lifeless and unconscious chemical elements. Time scales of millions, or even billions of years, of evolutionary processes are invoked to blur the improbability of properties emerging out of nothing, as has been extensively discussed in Chapter 4. These phenomenon are clear examples of the evolution principle, according to its adherents, whereby entirely new properties emerge out of nothing. That notion is wrong however, as we shall see.

The outcome of cause and effect is always predictable

Upon closer analysis, it will become clear that these processes, such as chemical reactions do not involve the creation of new properties out of nothing, but rather that they are manifestations of properties that were latently present. After all, if we were faced with the creation of entirely new properties, then this creation would have to be entirely unpredictable, and this is not the case. Bringing together hydrogen and oxygen under the same circumstances will always lead to the creation of water, without exception and totally predictable. Apparently, this process is about properties that already exist, but that transform themselves from potential to actual because of the changed circumstances. If they were completely new, meaning without cause, then they could be anything and therefore totally unpredictable. All chemical and physical reactions in which elements and properties are combined and which lead to new elements and new properties are always subject to the laws of physics and the laws of chemistry, and therefore, they are predictable. Even in quantum physics, where the behaviour of sub-atomic particles at a sub-atomic level seem to be unpredictable and arbitrary, the end result at an atomic level is always predictable and subject to the laws of chemistry and physics.

The different chemical elements in the periodic table are only a variety of combinations of atomic and sub-atomic particles. These combinations concern both the number of sub-atomic particles, for example the number of protons, neutrons, and electrons in relation to each other, as well as the geometric structure of atoms

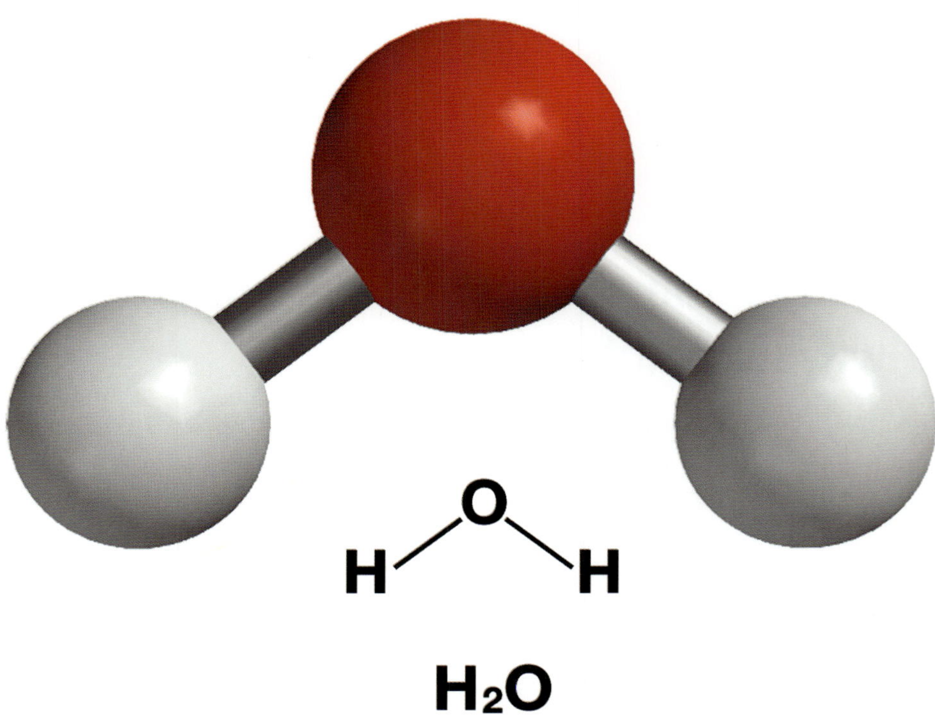

A water molecule, consisting of two hydrogen atoms (H2) and an oxygen atom (O). The properties of hydrogen and oxygen, both gases, differ a lot from water, a liquid, under the same temperature. Despite that this is a relatively simple molecule, water appears in various chemical variants. These variants come into being because of small changes in the atomic structure. The most well-known is deuterium, in other words, heavy water. Heavy water comes into being because the hydrogen atom, which normally has one proton in the nucleus with an electron in its shell, gets an extra neutron in the nucleus. This is also called an isotope, which means that the number of neutrons and protons in the nucleus are no longer identical. Hydrogen is an exception to this, for hydrogen does not have a neutron in its nucleus under normal conditions and only one proton. These small alterations can have major consequences for the properties of a chemical substance. Heavy water, for example, is used to curb nuclear reactions in nuclear reactors, a characteristic that normal water does not possess.

and molecules. Alterations in the geometric structure of a molecule can have major consequences for the chemical properties of a substance. The crucial point is that all chemical properties must potentially be present in the sub-atomic particles. The specific numbers, combinations, and structures cause these potential properties to be transformed into an actual property.

The transformation from potential to actual is no doubt a mysterious process. In order to make this somewhat understandable, one could compare it to the observation of a ball. Let's say that a ball has different colours on both sides; a red side and a blue side. An observer that looks at the ball from the blue side only sees the blue side and is under the impression that the entire ball is blue. Accordingly, when the ball is turned very quickly and imperceptibly to the observer it seems to have changed from blue to red. The observer is under the impression that the entire ball changed colour, since he can see only a part of the ball. In reality, the ball is both blue and red and both colours are simultaneously present.

We can conclude from the abovementioned examples that the most important argument in favour of the principle of causality is the notion that a specific cause leads to a specific, predictable effect. Such a connection is the embodiment of a certain law of nature. All scientific knowledge is based on this principle. However, if the effect is greater than the cause, or in other words, if there is no cause to an effect, then the predictability and the connection between cause and effect disappears. At that point science is no longer possible and the logical reasoning that is based on the same principle will no longer be in effect. In reality, we see a meaningful connection between cause and effect everywhere in nature as expressed in sciences, such as physics and chemistry. This again proves that the effect is always part of its cause, in reality or potentially.

The conclusion is that the evolution principle is being interpreted wrongly with respect to the difference between creation and transformation. What we witness everywhere in the universe is that apparently new developments are transformations of already existing but not yet perceivable properties and forces. This is the essence of the principle of causality and it is also the foundation and basis for the arguments which are the core of this book. This implies that the origin of the universe possesses all actual and potential forces and properties in itself and that the process of creation is factually a process of transformation. The principle of causality implies that all things that exist in the universe are explicitly or implicitly present in their cause. The material and efficient cause of the universe can therefore be regarded as an unbound field of existence, with infinite possibilities in which all possible and potential manifestations of existence are present.

The desire for simplicity

Science is based to a large extent on the method of reductionism, dissecting complex systems into its components, all the way down to its fundamental building blocks, the building blocks from which that complex systems derives its unique properties. Chemistry and the periodic table are good examples of this. Chemistry reduces complex, material chemical compounds into molecular structures, which in turn are comprised of atomic elements. The atomic elements are the smallest and fundamental units from which that element derives its properties and up until the beginning of the 20th century they were considered the fundamental and simple building blocks of matter. The discovery of sub-atomic particles such as protons, neutrons and electrons disrupted this vision and for a period of time electrons, protons and neutrons were promoted to the ultimate building blocks, even though their properties no longer corresponded to those of the atoms themselves. However, further research into particle physics, led to a completely different conclusion. The 'simple' building blocks appeared, in turn, to be extremely complex, consisting of numerous sub-subatomic particles such as fermions, hadrons, and bosons, and they are subdivided into quarks, leptons, baryons, mesons, photons, gluons and Higgs. Ultimately, according to quantum physics these sub-atomic and sub-sub-atomic particles turned out to be states of energy, in which complex and incomprehensible processes take place. The search for simplicity unravelled the sub-atomic universe into ever smaller particles and ever more complex structures, yet ending up in a mysterious, complex and energetic reality. The ultimate building block appears to be pure energy and while energy is an unambiguous term, and in that sense 'simple', it is at the same time extremely complex and full of actual and potential properties.

We can therefore wonder why simplicity is thought to be desirable and why it should be preferred in science. In other words, the desire for explainable simplicity is maybe desirable and also human, but not realistic. Einstein said:

"Everything should be made as simple as possible, but not simpler." [76]

It is therefore essential to give an explanation that is sufficient and not just simple. Reality is infinitely complex and permeated with properties that are also infinitely complex. Therefore, there is no reason to assume that the principle that should explain everything has to be simple. The only way ultimate simplicity can be achieved is by embracing nothingness as ultimate cause, and by assuming that existence has arisen, either partially or entirely from non-existence. Dawkins' assumption that the origin of existence has to be simple by definition and free of all complexity is incorrect and contradicts the very foundations of empirical science and logic. As stated already the assumption is based on a personal preference, a preference of an axiomatic nature, but an axiom that is contradicted by empirical and logical evidence. The universe indeed reveals a reality that is infinitely complex. Therefore, it is absolutely logical to assume that the cause of the universe is equally complex; a complexity that is eternal and timeless and that does not exist of anything else than itself. Based on these logical starting points, God, as an infinitely complex and timeless being, can certainly claim the position of an intelligent creator of the universe.

In every way simplicity simply fails as the ultimate cause of all causes and the ultimate explanatory principle. The singularity fails the simplicity test, with the big bang it gets worse, and once you end up with a universe there is nothing but complexity of mind boggling proportions. The moment that we talk about existence, whether or not we call this singularity, an exploding universe or the universe as we know it, we inevitably enter the area of complexity.

While early science was based on a very limited and explainable complexity, the scientific developments of the last 100 years have 'pulled-the-rug-out' from any such notion. Science has exposed a reality that contains an organized complexity of an unknown and inconceivable size. This complexity manifests itself from a 'simple' hydrogen atom to a living cell, and from a single living cell to a multi-cellular organism consisting of about 80 trillion cells, just like humans. If the origin of the universe is not simple but complex, then the question of where this complexity came from is basically irrelevant. This complexity is eternal and does not have an origin. It just is. Furthermore, if this complexity possesses the properties of consciousness and intelligence, then that also just is. If we accordingly decide to call this complexity God, with the addition that God just is, just like all things in the end just are, then this is a completely clear, plausible, and logical explanation.

Chapter 7

THE IMPERFECT UNIVERSE

"Is he prepared to prevent evil, but not able? Then he is impotent. Is he able, but not willing? Then he is malevolent. Is he both able and willing, then whence evil?"

David Hume in
'Dialogues Concerning Natural Religion'

"I don't know why we are here, but I am pretty sure that we are not here to enjoy ourselves."

Ludwig Wittgenstein

The fourth atheistic proposition – the problem of evil; the theodicy

The fourth proposition of atheism is, in the end, the strongest argument of all atheistic arguments. This proposition substantiates that the universe is imperfect taken from the human perspective. This imperfection is characterised by useless suffering, caused by the evil that people do onto each other, by the cruelty of nature, by natural disasters and the apparent uselessness of life in the face of death. Each person is in his life confronted with suffering and problems, sometimes caused by other people and sometimes caused by natural circumstances. Even in a perfectly peaceful society without war, injustice, and social inequality, there would still be suffering. Death, illnesses, the problems of old age; these are inseparable parts of life on earth as we know it. They are not caused by a political conspiracy, a bad government or a malevolent multinational, and even the CIA didn't do it. It is difficult to place suffering in a logical perspective. For suffering is confusing, it is the opposite of our desire for happiness, our most coveted emotion. What is of major importance here is the question of how we can link a universe that was designed to the existence of basic suffering that concerns all living creatures. Basic suffering means that each creature will at different times

Previous page: Fresco by Michelangelo from 1541, in which the Last Judgement is depicted in the Sixteenth Chapel of Vatican City. The Last Judgement is a theme that plays a major part in the Western religions, Judaism, Christianity, and Islam. The Last Judgement is the judgement of God on the individual acts of a person, weighing good and evil. If the individual soul did more bad than good, then it will go to hell. It represents the dualistic nature of this material world, where good and evil fight each other and where man is constantly tempted by evil.

in his life be faced with suffering. This does not mean that life only consists of suffering, for life in general is a combination of varying degrees of happiness and suffering. With regard to this, we can observe major differences in the quality of life of different individuals, how much happiness is bestowed on them or how much suffering they have to deal with. However, even the happiest person in the world will be faced with old age and death, and therefore his luck will run out in the end. Moreover, even the happiest person has to deal with the insecurity of possible suffering in the future, which is a form of suffering in its own right.

What is crucial here is the inevitability of death and the fact that each being will die eventually. The German philosopher Arthur Schopenhauer wrote:

"Human suffering only increases, for man actually know that death exists, while animals only fear death by instinct."

In his 2005 commencement address, Steve Jobs the CEO of Apple, after he had been successfully treated for cancer the first time, stated:

"No one wants to die. Even people who want to go to heaven don't want to die to get there. And yet death is the destination we all share."

For atheists the fundamental and inevitable suffering in the world is the strongest argument against the idea of a universe that was designed and against the existence of God, assuming that God is almighty and good. They state that power and goodness cannot be reconciled with suffering. The alternative is to degrade God to a being with limited power who apparently is not able to prevent suffering and evil. Even less attractive is the thought that God would not be good or merciful. For atheists it is more likely that God does not exist and that the universe was not designed. The core of the problem is that suffering and design seem to exclude each other and that it is difficult to reconcile them. Atheists however do not have a philosophical problem with suffering. They interpret suffering as a part of the necessary evolutionary processes that enable an organism to protect itself better against a hostile environment and improve its survival skills. For the suffering that does not have a protective function at all and does not serve any purpose, there is no explanation. But if there is no design and no all-embracing plan, then this explanation is not necessary anyway. Suffering is, according to the atheist vision, only a coincidental part of our existence.

Also, the universe, as it has presented itself to us, is chaotic and violent. Modern cosmology and astrophysics describe a universe that starts with a big bang, followed by billions of other explosions, supernovas, quasars, and all-devouring black holes. Our own earth is characterised by a lot of natural disasters, such as earthquakes, volcanic eruptions, floods, colliding meteorites, hurricanes and tsunamis. This is not exactly a harmonious environment suitable for man. One would expect that the universe created by a merciful, almighty and omniscient God would be harmonious, heavenly, and blissful. However, reality seems to be nothing like a heavenly creation. Looking at the quality of his creation, one would think that the current universe is a pilot project, a messy B-version and that God is still perfecting his creative skills. Or, as an alternative scenario, his rule is characterised by sadistic traits.

Evil: the strongest argument against theism

In philosophy the problem of the imperfect universe, or, in other words, the existence of 'evil', is also called the theodicy. The authoritative 'Oxford Companion to Philosophy',[77] calls this problem the most powerful argument against theism and the existence of God. An imperfect universe with suffering and evil does not agree with an existence of an almighty and merciful God. After the Second World War especially, and after the world found out about the horrendous atrocities of the Holocaust and the extermination camps, many even religiously inclined people started to doubt the existence of God. In the face of so much useless violence and suffering, such doubt is understandable. Many theological and philosophical discussions took place about the fact that God, should he exist, had allowed this inhuman and monstrous amount of suffering. Numerous publications appeared about this topic, under the heading of Holocaust theology where this difficult topic is discussed. The hell of the Holocaust—but also the other atrocities of the 20th century like the communistic terror administrations of Stalin, Mao, and Pol Pot, and crimes committed by various left or right wing dictators—have provided a new impulse to atheistic movements. Existentialism, materialism, nihilism, scepticism, and agnosticism are good examples of this. Human suffering as argument against the existence of God is so powerful, for it touches man in its deepest core. Intellectual thoughts and scientific arguments can be impressive, but they are often not able to deal with intensive human emotions.

The British philosopher and atheist William L. Rowe was a strong exponent of the problem of the theodicy. He concluded that the existence of evil in this world is a most powerful argument against the existence of God. Rowe stated the following:

1. *There are examples of intense human suffering that an almighty and omniscient being could have prevented, without a greater good having been lost, or without this leading to an even worse evil.*

2. *An omniscient, good being would avoid intense suffering, unless this could not be done without losing a greater good, or without admitting a greater evil.*

3. *Therefore, there is no almighty, omniscient, and good being.* [78]

Technically, this is a strong argument that is of course being disputed by various theistic philosophers. These discussions can

sometimes become quite technical, with arguments going back and forth. The discussion in this chapter on the problem of evil is limited to the points that are really important and that relate to our everyday experiences, whereby the arguments of William Rowe will be thoroughly examined.

Even though we conclude that the universe is extremely chaotic and violent and from a human perspective filled with pain and suffering, it is—at the same time and at the deepest level—extremely balanced and fine-tuned. This balanced reality is anything but chaotic, but it is filled with an infinite sophisticated complexity. The teleological and anthropic argument, the principle of cosmic fine tuning, as discussed in the previous chapters, do not leave room for another conclusions than that the universe most definitely is designed. So, there has to be an explanation regarding the problem of the theodicy.

The Vedanta philosophy

The topic of the theodicy is a philosophical topic by definition. While the first three atheistic propositions are primarily of a scientific nature, this fourth proposition is philosophical. It is a topic that focuses on the way man experiences the universe from an ethical and emotional perspective. Good and evil are not physical topics and they are not useful (or are hardly useful) for physical or mathematical interpretation. It is a topic that belongs to psychology, philosophy, and metaphysics. This does not imply that science does not play a part in such discussion. It is just a different part, a part in which science provides the tools and data based on which a philosophical interpretation can occur. This part of the book, from this chapter on, is therefore of a more philosophical nature. Considering the philosophical nature of this topic, other sources will be used than solely scientific sources in the area of physics, cosmology or mathematics.

One of these sources is the Vedanta doctrine, the core of the Vedic literature from ancient India. In former passages I already briefly referred to this 5000-year-old school of thought. The Vedanta doctrine is the philosophical foundation for the later development of two of the major world religions, Hinduism and Buddhism. While the Vedanta doctrine has a spiritual and metaphysical foundation, this doctrine is at the same time compatible with the insights and methods of modern science. The main reason for this is that the Vedanta doctrine explicitly acknowledges three sources of valid knowledge, namely sensory perception (called *pratyaksa* in Sanskrit), logic (*anumana*), and the mystical experience (*sabda*). This way, the Vedanta doctrine avoided the problem that many Western religions were faced with throughout the centuries. This is the conflict between religion and science, between spirituality and reason. This conflict has largely played a negative part in the development of Western culture. The Eastern religions, Hinduism and Buddhism, have had far less problems with this apparent contradiction. They had other problems, but the relationship between reason and spirituality was not a part of this. The Vedanta doctrine furthermore states that the mystical experience belongs to the category of sensory perceptions, but sensory perceptions of a higher order. The French philosopher Henry Bergson gave a beautiful example of this:

"In general it is assumed that the experiences of the great mystics are individual and extraordinary, and that they cannot be verified by common people, and that they therefore cannot be compared to scientific experiments, and that they cannot solve problems. A lot can be said about this. Firstly, it is incorrect to assume that a scientific experiment, or more in general, an observation that was determined by a scientific researcher, can always be repeated and verified. In the time that Central Africa was still 'terra incognita', the geographers were dependent on the reports of one single researcher, if his honesty and integrity was without question. The travel routes of Livingstone were placed on our maps for a long period of time. One could substantiate that verification is basically— and even factually—possible and that other travellers had the possibility to do research themselves, and that the map that was determined based on the results of one single traveller had a temporary nature until further research would make it definitive. I agree with this. But the mystic has also made a journey that others basically—or factually—can make. Everyone could do this, who had the courage of Stanley, who made a journey to find Livingstone. This is an understatement. Apart from the individuals who complete a journey at the mystic path, there are many who make this journey partially, and how many do you think are out there who only take a few paces, based on their own will or based on a natural tendency to do so." [79]

This quote of Bergson is very relevant. There are other realities that demand other instruments or senses to be perceived. In former chapters I have regularly referred to this issue. That also implies that additional instruments require additional efforts, not necessarily available to everyone. Scientists that wish to probe into the universe beyond the capabilities of our ordinary sense experience, have employed instruments with incredible capabilities. This enables them to perceive a reality not accessible to our ordinary senses. Which ordinary mortal would be able to employ a telescope like the Hubble, or the particle accelerator in CERN? Moreover, for non-insiders the observations generated by these sources are incomprehensible and they have to be interpreted by all sorts of derived technologies. The instruments that the mystic uses are again different from the technological wonder machines of the 20th and 21st century. The mystic uses the deeper sources of expanded and purified consciousness, whereby realities come into view that are beyond our normal senses and even our imagination. But this capability too requires additional efforts in order to be acquired, in this instance not money and technology, but inner development and expansion of consciousness. In Bergson's opinion this spiritual journey can be undertaken by any person with the sincere desire to discover this alternative spiritual reality using the same map and process used by previous explorers and seekers. For instance the long established and practised different systems of yoga, as practical extensions of Vedanta philosophy, are eligible and proven tools to facilitate such a spiritual journey. The reason I turn to the Vedanta doctrine is also because in this chapter we inevitably enter the realm of the metaphysical. The underlying assumption, however, is that the propositions used in this discussion, must be able to function as an extension of scientific insights and knowledge and, therefore, match that part of reality that does subject itself to scientific verification. Vedanta doctrine is well-suited to these requirements.

The attractive part of the Vedanta doctrine is that on the one hand it arose from the visions and the experiences of mystics over a period of many thousands of years. These mystics were practitioners of various yoga systems that are focused on the development and expansion of consciousness. On the other hand, the Vedanta doctrine is based on methods that can be compared to modern science, namely research and logic. The ancient system of logic which is part of the Vedanta doctrine and called *Nyaya* is well-known also among western scholars. Logical and critical analysis formed core aspects of the development and formulation of the Vedanta philosophy. The combination of mystical experience and critical logic has brought about a number of surprising insights. The most surprising insight is the fact that reality at its most fundamental level is paradoxical. This paradox comprises the relationships between energy and its source, between God and his emanations, between creator and his creation, for this relationship is both one and different at the same time. Because of this paradoxical relationship between oneness and difference, within the Vedanta tradition different schools came into being, where different thinkers either emphasized the aspect of unity or the aspect of multitude. In the early Middle Ages various Vedanta schools arose in India, varying from monism (oneness) to dualism (manyness and difference). Since simultaneous oneness and difference is paradoxical and therefore supra-logical, every school of thought —making use of the strict principles of logic— got stuck in the end with an unworkable system plagued by contradictions. In the 16th century these different interpretations were united in an all-embracing theory that is called *acintya bheda abheda tattva*. *Acintya bheda adheda tattya* literally means **'the principle of the incomprehensible, supra-rational simultaneous oneness and difference of reality'**. The founder of this theory was the great mystic and philosopher Caitanya Mahaprabhu who lived in India from 1486 to 1534. Caitanya united the various Vedanta schools and he laid the foundation for a new philosophical and scientific view of reality. In this book I often referred to the paradox and the principle of simultaneous oneness and difference. In the following chapters this principle is further elaborated on, specifically how it forms the foundation of the energetic reality and how it connects to the fundamental principles of the new physics.

Since we enter the realm of metaphysics at this point, the basic assumptions of the Vedanta doctrine are very useful to gain insight into the theodicy and the reality behind the material universe. Many of these insights match the insights of modern science. They are often the logical extension of these insights. Ultimately they also reveal the reality behind this universe; a reality that has been observed by other and higher forms of sensory perception. The impressive fact is that both realities are perfectly attuned. We have observed that also scientists like Dawkins, Rees, Hawking, Hoyle, and Einstein inevitable have turned to the area of philosophy and metaphysics. Modern science has no choice but to build theoretical, philosophical and logical models and constructs that exceed the limitations of immediately perceivable objects. Observation has become too abstract and too complex and is to a large extent dependent on a logical interpretation of that observation. Therefore, the criterion of truth has shifted from the immediate and verifiable observation to the logical, mathematical and internal consistency of a theoretical and philosophical model, as an extension of observation.

Use or abuse of free will

Just like in most theistic worldviews and religions, God is qualified in the Vedas as good, loving and merciful. The philosophical foundation for this qualification is simple; God is a fundamental unity, He is one with himself, and he consists of himself. The famous passage in the Bible where Moses asks God how to make himself known, is answered by God with the words "I Am who I Am" (Exodus 3:14). The translation from Hebrew has small variations of this text, varying from "I Am He who Is" up to "I Am who I will Be'. The core of this quote is that God is himself and cannot be reduced to something else. Which is apparently obvious, a matter-of-course, a tautology, and is at the same time a text with the deepest possible meaning. The meaning is that God is the Ultimate Being and cannot be reduced into something else. God is himself and consists of himself, and an object that is a fundamental and ultimate unity and that is one with itself can never be in disharmony with itself. From an ontological and 'being' perspective of reality this is logical. If we project this principle onto a state of being related to consciousness and emotions, then the same logic is applicable; with the difference that we can replace the words 'unity' and 'harmony'—in an ontological context—with the word 'love' in the psychological context. Therefore, the terms unity, goodness, harmony and love are exchangeable when we talk about the qualifications of God.

Evil and wickedness, that which is bad and malign, is based on disharmony, the opposite of harmony. Within the nature of God such an internal conflict is impossible. He is therefore in harmony with himself and therefore Good. This basically is also true for us as living creatures and as one of his numerous energies. His energies are also part of Him and therefore in harmony with Him. However, as living and conscious creatures we have a limited free will and a degree of independence. While free will is a core attribute of consciousness and a glorious manifestation of our eternal individuality, it can also be a source of complication that can bring us in conflict with ourselves, with others, with the universe and ultimately with God himself.

The Vedic doctrine states that the material universe is only part of the total creation of God. God has unlimited energies, which can be sub-divided into two main categories, matter and spirit. These two categories of energy are qualitatively different, in that matter is defined as an inferior energy, and spirit as superior. Each energy forms the substratum and bases from which the different types of realities are constructed. From matter the material universe is constructed, in which we presently reside. From spirit the spiritual universe is constructed, which is, while made of superior energy, not shapeless or without attributes, but actually possesses attributes of an incomparable and superior quality. The spiritual universe indeed meets all expectations of what an almighty, omniscient and merciful creator would have accomplished. The spiritual universe is permeated with a heavenly harmony and is characterised by the total absence of pain, grief, fear, and death. It is a universe that according to the Vedas, is ruled by three principles: sat (eternal existence), chit (full knowledge), and Ananda (permanent and unlimited bliss). There is a third category of energy, which constitutes living beings such as ourselves. In essence living beings are eternal atomic sparks of spiritual energy, and as such belong to the category of spirit. However, due to their atomic size, living entities may become associated with the material energy, and in fact become conditioned by this energy. The reason for such conditioning and the effect thereof will be discussed later on in more detail.

If such a spiritual universe indeed exists, then the logical question would be why we are not there but here; here being the material universe, the world of birth and death, of black holes and exploding supernovas, and a world where the existence of God is largely hidden and invisible. The Vedas explain this dilemma as follows. Each living being is in its core, on the one hand, a conscious, independent and worthy individual, but on the other hand a dependent part of God. As part of God we possess the same qualitative attributes, namely *sat, chit,* and *Ananda,* but we are quantitatively different. God is spatially infinite and omnipresent, while we are spatially finite and localised. As a part we are inextricably connected to the whole and totally dependent on the whole. Nevertheless, we possess— as a part—also a conscious individuality, with a certain amount of independence. This independence manifests itself as conscious free will, in which we can make the fundamental choice to function as part in harmony with the whole, or to try to enlarge our independence. A disproportionate desire for independence can lead to disharmony and bring us into conflict with the cosmic order. In the end, it can also bring us into conflict with God, the creator and maintainer of this cosmic order. This conflict can even take the shape in which we 'wish' the presence of God 'away'. The material universe is created for these creatures that desire a larger amount of independence, in various gradations. In a good and benevolent manner this is expressed in desire for power, but with the intention that this power is used in service of the cosmic order and in service of God. In a less innocent and less benevolent form, this power can lead to a corruption of the soul and the desire for self-glorification. Egotism and narcissism are the symptoms of a corrupted soul that cannot accept its subordination. What started out as a desire for greater independence and more power can end up as an actual desire to be God. Psychologists believe that acquiring power only leads to the desire for more power. The desire for power is almost by definition a desire that can never be satisfied and leads the soul in the direction of self-destruction and evil.

The material universe was purposely created in such a way that the hand of the intelligent designer is nearly invisible and where it indeed seems as if chaos and coincidence are the leading forces. Therefore the living beings in the material universe develop the illusion that God does not exist and that they themselves represent the highest form of intelligence in the universe. This illusion is called *Maya* in the Vedas. The core of *Maya* is that it constructs reality in such a way that it accommodates the desire of living beings to, on the one hand, enlarge their independence from God and on the other hand to actually try to become God. In other words: the entire material universe is constructed around our own desires. Many of us probably do not recognise or identify with such ambitious desires. However, tendencies often only become well recognisable when they are presented in an extreme form. The extreme examples of the sick desire for power is something we know all too well from history and many of us have experienced this personally. Hitler and Stalin and the atrocities of the Second World War have shown how the human mind can degenerate to an extreme desire for power and self-glorification and to total and demonic degradation. The majority of people will not recognise themselves in the sick minds of people who caused the Second World War, the Holocaust, and the Gulag. Most of us are contaminated with lighter forms of narcissism and egotism that are expressed in small idiosyncrasies, jealousy, aggressiveness, and frustration. Furthermore, we have to place our current condition in the context of a past that does not only limit itself to our activities in this life. Our existence in

this material universe is, according to the Vedas, ruled by the laws of karma and reincarnation.

The law of reincarnation

The cornerstone of the law of reincarnation is the fact that individual consciousness is a fundamental, eternal energy that is irreducible and exists in its own right. Consciousness is not a by-product of matter or an epi-phenomena of the brain. Consciousness is an energetic reality, that is, while expressing itself through brain, nerves and body, in the end not determined or limited by matter. Consciousness is non-material and possesses an eternal identity that is called *atma* in the Vedas, meaning soul. The soul resides in the physical body just as a driver sits in his car. As long as the driver sits in the car, he will operate the engine by means of instruments, switches and pedals. However, without a driver the car will not move an inch. Similarly the soul operates the body, where instruments, switches and pedals are replaced by senses, mental processes, brains and the nervous system. Without a soul, however, these processes and instruments are literally dead. The soul generates two essential forces: consciousness and vitality. These can be subdivided into intelligence, emotion, desire, and will-power. Together they form the attributes that we call intelligent life. A crucial fact is that the soul, just like the

In previous page, a beautiful illustration of the reincarnation process. According to the reincarnation doctrine each person, each living being, is a spiritual spark where the "I" and consciousness are located. This spark, soul, or atma (Sanskrit) is the self that is located in a material vessel, the body. Reincarnation does not just take place at the time of death. Already during his lifetime the soul reincarnates gradually, for the body changes from baby, to childhood, to adult to old-age. In this process of change, each part, each cell and each atom of the human body is fully replaced—several times—by entirely new building blocks. The soul, the source of consciousness and Self-awareness, is the permanent factor in this process of change. Death means the end of the process of gradual reincarnation. On the moment of death, the soul leaves the body suddenly and gets a new body based on the laws of karma. The continuation of the reincarnation process is ultimately determined by our own desires, our attachment to the mortal body and the way we act in pursuit of our desires and ambitions. Changing these desires and giving up our attachment to the mortal body are important steps to break the cycle of birth and death. The various yoga systems were designed to give man the opportunity to escape this endless cycle of birth, death and rebirth.

driver of the car, differs from the vehicle, the body. Just like a worn-out car has to be replaced by a new car, the same way an old body needs to be replaced by a new body. This process is called reincarnation. In the Bhagavad-gita, the famous 5000-year-old text from India that contains the essence of the Vedic knowledge, the process of reincarnation is described beautifully as follows (chapter 2, verse 13):

"As the embodied soul continuously passes, in this body, from boyhood to youth to old age, the soul similarly passes into another body at death. A sober person is not bewildered by such a change." [80]

According to the doctrine of reincarnation, the body is only a temporary, material vehicle the soul makes use of during its journey through the material universe. As long as the soul is attached to the physical body and to the physical world, it will continuously be reborn in a new body. The nature and quality of the new body is determined by karma that the soul generates throughout its stay in its previous body. This chain of birth and rebirth is called *Samsara* in the Vedas; the cycle of birth and death. In the end, the attachment to the body is determined by the previously mentioned basic attitude in our relationship to God. This attitude is determined by service and harmony on the one hand, and jealously and a desire for independence on the other hand. This is the spectrum within which our relationship with God, the relationship between the part and the whole, is in the end defined by ourselves.

The laws of karma

The law of karma is linked to the law of reincarnation. This law arranges the continuum of our actions and the consequences of our actions beyond the limitations of birth and death. The law of karma determines that each action we take will evoke an equal reaction in the future. The contents and quality of that reaction is determined by the contents and quality of our action. If we do good, then we can expect good in return in the future. If we act evil and cause unnecessary suffering onto other creatures, then we will have to suffer ourselves in the future. Of course, this makes our lives extra complex, since on the one hand we have to endure the reactions, good or bad, from our actions of our past lives. On the other hand we undertake new actions that are partly determined by karmic responses, but are also partly determined by our free choices.

At this point it is a good to analyse what we consider to be good or bad and how we experience good and bad. Intuitively we all know this: a good person has respect for others and will not unnecessarily harm another being. A good person likes others to be happy and will never pursue his own happiness at the expense of others. A bad person is basically the opposite. He is self-righteous and prepared to cause suffering onto others in the pursuit of his own happiness. Even worse is he who enjoys the suffering of others. The essence of the law of karma is maintaining universal order. The universe consists of a multitude of parts that are part of an organic unity. The parts have to be prepared to serve the whole and to conform themselves to this. The rule of law and the social order of human society is based on the same principles and is a reflection of the laws of karma. Each individual has to know his place in this social and legal order. If someone allows himself more freedom and transgresses the rule of law then he is an offender, or worse, a criminal, depending on the severity of the transgression. The core of our human legal order is that our freedom should not lead to an unnecessary harming and damaging of the interests of another person. The essence of the law of karma is therefore that we are not permitted to cause unnecessary harm to others and moreover that we should try to help others, whenever possible, to avert suffering and increase their happiness. This is an essential distinction between good and evil.

The laws of karma and reincarnation provide viable explanations for many of the complexities and dilemma's we experience in life, specifically from a human psychological, moral and emotional perspective. That does not mean that the laws of karma themselves are easily understood and that they are not extremely complex themselves, for they are indeed. First of all there is the dilemma that we cannot remember past lives, except for those who may have had paranormal experiences and remembrances of past lives. Much scientific research in the field of psychology and para-psychology has been conducted for over hundred years and much evidence has been accumulated that experiences of past lives can be real, and that reincarnation does exist. Yet a vast majority of people will not have had such experiences, and in any case the question remains why we cannot remember past lives. Vedanta doctrine explains that our desire to enjoy independently from God requires that God's existence becomes largely veiled and obscured. It also requires that we identify with our body and that we forget our real spiritual identity as eternal spiritual beings. This identification with our bodily identity requires ignorance and forgetfulness, for instance of our past lives. In that state of ignorance we assume a false bodily identity whereby we are acting out various roles according to our karma, birth after birth.

When we go to a movie theatre to enjoy a good movie, we voluntarily enter a state of ignorance and allow ourselves to be submerged into a bubble. First of all the lights are switched of whereby the darkness makes us forget our real environment. The movie itself is obviously a construction of actors and effects, staged in the movie studio and recorded on tape or any other recording device. The projection of light images on a screen is similarly an illusion cleverly orchestrated to create the appearance of reality. All these arrangements are created just so that we can identify with the movie, which identification enables us to experience emotions of fear, sadness or happiness depending on the storyline and the accompanying images projected on the screen. If any one of the components that keep us in ignorance during the movie experience would fail, for instance lights that accidentally switch on during the movie, it would destroy our ability to actually enjoy the movie. Imagine trying to enjoy the movie if we can see the entire set, the green backdrop with images transposed or the city created of cardboard, or the characters stopping every few minutes for a break and reverting to their real personality, before getting back into character. It would simply be impossible to enjoy. Similarly, our desire to enjoy separately from God is impossible without creating a believable illusion that enables us to identify with our body and physical environment, and enables us to believe that God does not exist.

Another problem is the sheer complexity of the process of action and reaction. Every individual creates numerous reactions, good or bad, on a daily bases to be received at some point in the future. When we imagine that this is going on with countless, infinite living beings, where significant numbers are interacting with each other in different ways since time immemorial, then it is easy to understand the complexity involved in connecting each action with an equivalent reaction that fits within the specific circumstances of each individual across space and time. It seems extra-ordinary that our lives would be subjected to such extreme mechanisms of registration and control, spanning lifetimes since time immemorial. On the other hand we can observe that our modern digital society has enabled governments to exercise control over its citizens that would have been unthinkable prior to the advent of the micro-chip and the internet. While they are not able to exercise such control across lifetimes, what they are able to do within the lifetime of one individual is most impressive. God's all-pervasive intelligence and control extends to every atom and sub-atomic particle, therefore it is no surprise he is able to do so in for instance human affairs. Thereby it must be remembered that such control and recording capability can in the end become overruled by the exercise of free will on the part of the living being.

Evil cannot be understood without karma and reincarnation

The problem that Western theologians and philosophers are faced with today is that without the understanding of karma and reincarnation much suffering seems to be pointless and arbitrary. The cited arguments of William Rowe solely refer to this. Rowe emphasises the apparent uselessness of the suffering and the fact that an almighty God—if he would exist—does not stop it. Of course, without the understanding of the laws of karma and reincarnation, much suffering is inexplicable and therefore apparently useless. If suffering is reduced to the personal responsibility of the individual and not to the unfathomable arbitrariness of God, then 99% of the theodicy is solved. The laws of karma and reincarnation are very explicit about this. The suffering we experience is caused by the unnecessary suffering that we have inflicted on others in the past. In that sense, all suffering is justified and the result of voluntary choices that we made ourselves in the past. This proposition can only be defended, however, if we accept the principles of karma and reincarnation. The complexity of our current life and the major individual differences between people in terms of happiness and suffering can of course never be explained based on one single life. This is a major dilemma for the Western religions that have rejected the principle of reincarnation. This weighs heavy on the apparent arbitrariness of God.

Many of the arguments that philosophers and theologians use against each other boil down to this problem. Regarding all discussions concerning the cause of evil, the argument of free will is, in the end, the only argument that really holds out. On the one hand, it is evident that consciousness and personality cannot exist without free will. On the other hand, because free will is the only instrument that can disturb the natural, internal harmony between the part and the whole. The problem is that in a one-life-only scenario there is a lot of individual suffering that cannot be retraced to my free will and individual choices. The explanation regarding this type of suffering varies from necessary suffering to get to a greater good or necessary suffering to avoid a greater evil. These arguments, in the end, can be led back to the principle that God, of all possible creations, made the best possible creation, which is the universe we now live in. This is an argument that was among others propagated by the German philosopher Leibniz (1645-1716). This implies that all evil that is not generated by individual choices is necessary evil, without which creation could not function. On top of that, Leibniz assumed that the happiness of living beings in this universe was not the only goal of God. Evil also has a functional role to let man grow and to shape him.

This shows, with all respect for the person who came up with this thought, not a lot of respect for the creative, almighty qualities of God. When we now look at the unimaginable complexity of this material universe, then we have no choice but to conclude that the creative ability and intelligence of God are indeed unlimited. This also implies that the results of the creation could have been totally different, if God would have wanted it.

For example, look at the process of mitosis or cell division and the fact that the human body still grows old. Science does not understand why the regenerative ability of cells cannot go on forever, so a person can become hundreds of years old, or thousands or millions, or even can live eternally in this material body. Somewhere in the DNA of each cell there is, up to today, an invisible instruction that negatively influences the division of cells after a certain period. Regardless of whether you would want to live forever in a body that is still vulnerable and can still experience pain. However, also the experience of pain and the balance between pain and pleasure seem to be arbitrary, a balance that is not in our favour, Certain organs can generate limited bodily pleasure, but these same organs can cause us a great deal of pain. For example, look at our ability to enjoy food; every day, three times a day, at an average of half an hour. However, a cancer patient that suffers from stomach cancer or cancer to the gullet may suffer great pain every day for years. For this there is also not a clear explanation; that balance could easily have been more favourable to us human beings.

This universe is not the best possible creation

The Vedic philosophy is clear about this. This material universe and this planet earth are not, by any stretch the best possible creation of God. On the contrary, in the Vedic doctrine the material energy, and I just referred to this, is qualified as an inferior energy and the material universe is regarded as an inferior creation. There is a somewhat comic remark about this that states the difference between an optimist and a pessimist:

"According to the optimist this world is the best possible world out of all possible worlds. The pessimist fears that this could be true."

The material universe with all its different levels of existence is created and designed around our desires and around our karma. As such, the quality of creation, at least the part of creation that we live in with its limitations and possibilities and with its joy and grief and the balance between them, is determined by ourselves and is the result of our own individual and collective karma.

Our journey through the material universe is a learning process, in which we have to gain insight into the fact that if we desire too strongly to be independent of God, this will—in the end—lead to conflict and frustration. This realization linked to the insight that we should fulfil a serving role at all times can lead to a fundamental change of consciousness and attitude. In the end, this change can lead to a liberation from the material world and prepare us for a return to the spiritual universe. An important part of this change is the development of the realization that God exists, linked to an appreciation and admiration for Him. Different yoga disciplines use techniques that, for example, help to develop a cosmic consciousness and, accordingly, a God consciousness. When making use of these yoga techniques, amongst which are meditation and certain internal cleansing processes, then this in time can lead to the development of feelings of love for God. These feelings of love are our natural condition of being, since we, as was remarked before, are a part of God. And a part is, in an ontological perspective, always in harmony with the whole. In a psychological, emotional context the word harmony can be replaced by the word love. In other words, the individual soul, as conscious part of God has a loving connection to God and in turn God has a loving connection to the individual soul. The love is generated by the conscious observation of the living being that God is infinitely full and infinitely attractive. The things that stroke our senses in this world, or the people we enjoy, have certain attributes that we find attractive. Beauty, intelligence, power, charm, harmonious tastes, sounds or smells; these are attributes that we like, or that enchant us or even make us passionate. According to the Vedas, God possesses all attractive attributes in an infinite amount. He is described, with regard to this in Sanskrit as *Bhagavan*, the all-attractive one and He who is infinitely full of all opulences. More specifically, *Bhagavan* is defined as He who possesses all of the six opulences in an infinite amount:

1. **Beauty**
2. **Intelligence and knowledge**
3. **Detachment**
4. **Fame**
5. **Wealth**
6. **Power**

In this spectrum of beautiful attributes, the observer of these attributes becomes enchanted and enamored. Just like a man may fall in love with a beautiful woman, and vice versa, a living being may fall in love with and be enchanted by the Supreme Being. The impressions and sensations are overwhelming. With regard to this, certain attributes can make a greater impression than other attributes. All attributes of God are enchanting and attractive, but there are nevertheless important differences. The first three attributes mainly refer to the internal qualities of God, his own intrinsic personality. Beauty is, with regard to this, the most determining and all-embracing factor, since this directly speaks to the emotions of an observer of these attributes. The second three attributes reflect his energies, his power, and his riches as they are expressed in the infinite and majestic creation. Of these last three, especially strength and power are the most determining. Fame and riches arise from the infinite power of God. Beauty and power are therefore the two extremes in this spectrum of divine qualities.

Love and power

The perception of unlimited beauty, intelligence, kindness, compassion and detachment in another person will generate a feeling of admiration and especially love. This love is totally focused on the other, the possessor of these enchanting attributes. If a living being is enchanted by the beauty and love of God, then he will focus completely on the object of his love and he forgets himself. Love is an emotion that is in its purest form unselfish and not focused on the "I". A living being can also be enchanted by the overwhelming power that God radiates. Also in this enchantment he can lose himself. There is, however, a subtle difference between the two forms of attraction. The subtle difference between the two attributes is that beauty is part of the internal attributes of God—it reflects and reveals his own identity. Power, on the other hand, is focused outward and it reveals his energies, his energetic ability. Power implies duality, the relationship between the possessor of power and that which is subjected to this power. Inevitably the attraction to power leads to a focus on the I, for the focus on the I is enclosed in power. In the end, the attraction to the energy of God can be stronger than the attraction to God himself. The increasing focus on the I can even lead to the desire to possess these energies oneself and to control them. The actual possessor and controller of these energies is ignored with regard to this by the living being, and they become unjustly appropriated. It is this desire for power that is the foundation of the corruption of the living being. In the end, it is the choice that a living being can make in his eternal existence and his eternal journey. The choice between love and surrender on the one hand, or the choice for strength and power on the other hand. In the beginning of this development the differences are subtle, but later on they become striking and they will lead the living being in two fundamentally different directions. While in both cases the emotional affection will be focused on God and his divine attributes, the one emotion leads to a non-focus on the I which is love for God, and the other emotion leads to a focus on the I, which is love for God's energies. Therefore, put differently the choice of the living being is determined by love for power on the one hand and the power of love on the other hand.

The possibility to make choices, to have the ability to be drawn to the different aspects and attributes of God is an intrinsic feature of consciousness. Consciousness without the multitude of observations, without the choices of observation and without the choice to act in accordance with our observation, does not qualify as consciousness. This is in the end the explanation why there are different worlds and why different living beings make different choices. Consciousness without freedom is not consciousness, but freedom also brings about risks. The material, physical world is the result of such a risk, with regard to which at a certain point in time in our eternal history, we chose power above love, masterdom over servitude. In turn, the desire for power leads us on the path of evil, whereby the boundaries of what is rightfully ours and what is not, starts to fade. In this process we cause suffering onto others and thereby we create the foundation for our own suffering. We can however control our desire for power and adjust our ego to the right proportion again. The choice we once made is a choice we can revoke, at any time and that is also our free will.

Imperfection, wonder and our search for knowledge

However, in order to be able to make this choice, we first have to become aware of the fact that there is a choice at all, and we also have to become aware of the nature and contents of this choice. The problem is that our existence in the material universe is characterised by the fact that we are immersed in illusion, or in Vedic terms, in *Maya*. We are fundamentally deprived of knowledge. That implies among other things that we are not aware of the fact that the imperfections of the universe are structural and designed. We experience these imperfections and the resultant suffering, but we do not have a logical explanation for them. Because of our ignorant condition, we tend to come up with all kinds of alternative, sometimes delusional theories that often worsen our position. Acquiring knowledge, however, demands a serious effort, and in order to make this effort, we have to have a motivation and a desire.

According to Vedanta doctrine there are two sources that may motivate us to acquire knowledge. Firstly, as explained previously, every living being is a spiritual and eternal soul, and at its core intrinsically *sat-chit-ananda*, eternal, full of knowledge and blissful. Therefore every soul has a natural and intrinsic desire to regain its original state of being, which explains the desire for knowledge. Secondly, this desire is, for the larger part, fed by the realisation that our existence in this world is imperfect. As we are born into this world we are not immediately aware of this imperfection, and in many ways the realisation of this imperfection is part of the process of growing up and becoming an adult. Looking back at my own childhood and adolescence, the biggest shock was the realisation that the world was not perfect. Something that has always amazed me about the childhood experience is the fact that you experience life and existence as a self-evident given. In this state of mind, you experience life as something that is the way it is, that has to be the way it is, without really questioning it, which makes it implicitly perfect. This changes as you grow older and you start to realise that life is not perfect and slowly or suddenly the world loses its innocence. You realise that your parents are not perfect, that the world is full of problems, that people do absurd things such as fighting useless wars, and that human history is an embarrassing chain of human conflicts and cruelty. You realise that life is finite, imperfect and full of limitations. Emotionally these limitations are expressed in different degrees, from a lack of satisfaction to more serious forms of suffering. While we desire happiness, we find that this goal is elusive and accompanied by obstacles and problems, a painful and unpleasant discovery. At an intellectual and cognitive level, limitation expresses itself as ignorance, with an additional complication that ignorance includes the ignorance of one's own ignorance. Yet, when applying sufficient introspection and honesty, we inevitably come to realise that our understanding of the world and of life is indeed very limited. Just as we long for happiness, we also long for knowledge. This longing makes us inquisitive; we just want to know because we want to know. We also want to know because we have to act and make decisions. In this respect, a lack of knowledge can lead to frustration, when our ignorance leads to undesired results. We spend the first 20 years or so of our lives in schools and educational institutions just to combat ignorance. Ignorance, as history teaches us, can have disastrous consequences, varying from eating a poisonous mushroom to the collective ignorance in voluntarily choosing Adolf Hitler to be the leader of a country. Finally, we want to know because we are confronted with the imperfections of life, such as old age, disease and death accompanied by worries and uncertainty. We want to know why we have to suffer, how to deal with that suffering and especially how we can prevent future suffering. Because of these limitations and imperfections, existence ceases to be a self-evident and matter-of-course experience. And when something is no longer a matter-of-course reality, then it automatically becomes an object of wonder and incredulity and causes us to investigate our existence in a search for an explanation.

Knowledge and action

Since we live in an imperfect universe with all sorts of imperfections and limitations, everyone is concerned about his future and wants to know his fate. We want to know what is in store for us amongst all the joy, suffering, happiness, and unhappiness around us. We also want to know how we have to behave to increase our happiness and to avoid misfortune. In the end, we try consciously or unconsciously to understand the laws of karma and to act in accordance with these laws. Intuitively, we all know that our actions have consequences and that there are causal links between the things that happen to us and our actions. Some links are simple and obvious: when I park my car in the wrong place, I will get a parking fine. When I eat too much, I get fat, and if I rob a bank there is a strong probability I will end up in jail. There are, however, also actions and reactions of a subtler nature with consequences that manifest themselves at a later time. Moreover, we are continuously confronted with the reactions of our actions from the past, both from this life as well as from our previous lives. This makes our life complex and oftentimes confusing. Human beings are not only active by nature but they are constantly forced—because of the circumstances of our earthly existence—to take action. We have to feed ourselves, dress ourselves and protect ourselves and we have to take care of our family and children. Our entire life we have to take action and we cannot not take action. We have to act; but to take action, no matter how small the action, we have to make a decision first. We make that decision based on acquired knowledge, our worldview, and our philosophy of life. Our decision will depend on what we consider to be good and advantageous, or what is bad and disadvantageous. The question thereby arises whether life is just a series of pre-determined events that happen in a deterministic manner, or whether human beings have a degree of free will and are able to make their own choices that will shape their future. Generally, people are inclined to believe that we have a free will, which they sense as a self-evident fact, and that therefore everything we do matters. However, some mechanistic, materialistic theories claim that our consciousness, our thoughts, feelings, desires, and will power are only the result of a number of pre-determined physical and chemical reactions. The Vedanta doctrine is clear about this, human beings do have a free will and individual choices can make a difference. Yet past karma has a major influence so that our lives are effectively mixtures of predetermined events and free choices, which of course adds to the complexity of life. These and similar related questions are of vital importance when determining how to live and how we make choices in this complicated world. We are looking for answers so we can make the best and proper choices in life. These answers are part of the bigger answer, the answers that explain our existence and that provide us with insight into the bigger picture, the universe we are part of and our relationship with God. As we have stated in the beginning of this book, the question whether or not God exists is the most important question that a person can ask himself. The answer to this question determines all other questions in life and also all accompanying answers. The imperfections in this world and the suffering we all have to go through in the end, make us aware of the fact that something is not right and that we do not belong here. These are in the end the most important motives behind our search for knowledge.

"Why does the universe go through all the bother of existing?"

Stephen Hawking in 'A Brief History of Time' [81]

Non-existence does not exist

At this point it is important to revisit the philosophical question that was presented in the first chapter: why does existence exist, why is there something rather than nothing? This is a question that has been on the minds of philosophers and theologians for thousands of years. The question has been revived by the discovery of the Big Bang and its presumed preceding condition, the singularity. Big Bang Cosmology for the first time in the history of science determined that the universe had a beginning, that it was somehow created, and that there must have been an origin and cause to the universe. The standard theory is that the Big Bang came out of the singularity, an infinitesimal almost non-existent point in which all of the energy in the universe was compressed at an infinite density. The question that then follows immediately is where did the singularity come from, was it eternal, or did it have an origin of its own. In answering that question today an impressive number of scientist believe that the singularity and the subsequent universe arose from nothingness. The famous and brilliant physicist, cosmologist and mathematician Stephen Hawking holds this view, even though he is sometimes somewhat vague in his statements concerning

In previous page, billions of stars: it is impossible to count them in this picture. This is an astrophoto of a spherical cluster named 47 Tucana, an object in the southern hemisphere. Each light point is a star and a sun, and they all move towards an enormous black hole in the centre. The centre looks like a white bubble, for the number of stars is too large to be perceived separately. This photo was made by a telescope in Australia. The universe is unimaginable in every way and it is infinite and incredibly complex. However, all this would have originated out of nothing, without cause, necessity or reason. An appropriate question to wonder is why nothingness would undertake such an enterprise;

this topic. In a way, it sounds attractive of course, especially when it is presented in a scientific jargon: 'Nothingness is the Ultimate Cause of Everything'. It is a way to be mystical and to even sound religious without being religious. It is a way to delve deep into the ultimate mystery of existence without succumbing to an overly devotional humility. It rather is a proud humility of a brilliant intellect that in the face of the infinite Everything, knows how to reduce this to a mysterious, but controllable, Nothingness. Despite the obstacles and impossibilities presented by physics, many cosmologists and physicists actually believe that the singularity and the whole universe arose from nothing. Therefore, the how and why the universe came into being out of nothing is a very relevant matter in science today.

At the same time, as we shall see, nothingness is being used, or rather abused by atheist thinkers as an escape from a major, insurmountable philosophical, logical and scientific trap. Nothingness as ultimate explanation contains a massive scientific and logical contradiction that fuels a key argument against atheism, which argument is centred on the relationship between being and none-being and the none-existence of nothingness. The argument that's being put forward here, for this purpose labelled the Fallacy of Non-Existence, states that atheist philosophy relies on the existence of non-existence in order for their line of reasoning to make any sense. This reliance is not explicit and direct but rather it is hidden and veiled by clever logical constructs. The crucial flaw is that reliance on something that is non-existent inevitably leads to logical contradictions and logical impossibilities of the square circle type. The Fallacy of Non-Existence as argument against atheism is being put forward in addition to the previously presented arguments, such as the teleological argument and the argument that refers to the defects of materialism. It is an argument that is also closely related to the atheist proposition that the ultimate cause of everything must be of ultimate simplicity, and the refutation of this argument in chapter 6. To some extent, the Fallacy of Non-Existence overlaps with the arguments presented in chapter 6, yet in another way it elaborates on them and approaches the dilemma from a different angle. Chapter 6 dealt extensively with the argument of simplicity, the laws of cause and effect and the laws of logic, that contradict and refute this fallacy. Implicit in these arguments is the role of nothingness and non-being. In this chapter this role is explicitly expanded upon, specifically explaining how non-being, because it does not exist, by definition cannot play any role in existence. This is also precisely the definition of the Fallacy of Non-Existence: Non-being, none Existence or Nothingness do not exist at all, not even just a little bit like the singularity.

The core of the argument against the Fallacy of Non-Existence consists of three propositions, all three of which equally express and defy the absurd reality they refer to. Since the first proposition forms the bases of the refutation of this fallacy, the following two only increase the absurdity of what they refer to. Nonetheless we will walk through them and dissect them, because the very thing that is being refuted by these propositions is being employed by atheist thinkers as factual and truthful. This is done in an attempt to answer one of the most puzzling and startling question that is facing science today: what is the origin of the singularity and what caused it come into its, however minute, existence. These are the three propositions:

1. *Nothingness, or non-existence does not exist, in the most literal sense of the word;*

2. *Since nothingness does not exist, it cannot produce anything, so therefore it cannot produce something such as the universe;*

3. *If however Nothingness were able to produce something, then it would be totally illogical and inefficient to do so.*

Absolute nothingness and relative nothingness

The first proposition, 'nothingness or non-existence does not exist' is apparently simple and obvious, a tautology like stating that 'a door is a door' or 'a cat is a cat', but upon closer examination it is not that simple. After all, we talk about nothingness, as in this discussion and we sort of understand what is meant with the term 'nothingness'. Despite the fact that nothingness does not exist, it does have a meaning to us. The reason for this is that we have to make a distinction between nothingness in an absolute sense and nothingness in a relative sense. Relatively speaking, we use the term nothing all the time in our everyday lives. I can refer to an event that did not take place: "…nothing happened." I can also refer to the fact that Tom did not show up for a meeting, so: "…Tom did not show up." There are many situations where we talk negatively about the presence or absence of something or someone. The number zero, as it appears in mathematics, is also a relative notion. After all, zero exists because of and in relation to the other numbers.

Absolute nothingness, on the other hand, has an entirely different meaning; at least those who use the term in that sense have a different meaning in mind. Absolute nothingness refers to real nothingness, to non-existence, the total absence of anything and everything. Absolute nothingness therefore does not exist, for that is the exact definition of the term 'nothingness'. Nothingness, non-existence, does not exist. We cannot, therefore, imagine absolute nothingness, except for a 'relative' notion of it.

In *Les Misérables* Victor Hugo made the following, in some ways obvious and self-evident observation:

"Nihilism does not have a substance. There does not exists something that is 'nothing' and zero does not exist. Everything is something and nothing is nothing." [82]

The existence of non-existence is not only a paradoxical suggestion, it is also an impossible suggestion, comparable to the suggestion that square circles exist. Everything that I know about non-existence and everything that I can imagine about it comes from something that exists: my consciousness. I exist, and my consciousness exists, so my ideas of non-existence take place in the world that exists. This creates a fundamental paradox. It can be compared to the proposition of the famous French philosopher René Descartes when he stated, "I think, therefore I am." Doubting our thoughts is a confirmation of our thoughts in the same way that doubting existence or contemplating non-existence is a confirmation of existence. This leaves little or no room for the existence of absolute non-existence. Perhaps the closest possible notion of nothingness would be the void, completely empty and infinite space. But even absolute and completely empty space is not totally nothing, strictly speaking, for space still has dimensions even if these dimensions are of an infinite nature.

Atheist philosophies employ the term 'nothingness' in the absolute sense, which follows from their notion of a pluralistic universe. In atheism, existence is only an incidental by-product of eternal and infinite nothingness. 'Being', in the form of a universe for instance, is optional —it can be there and sometimes it is not there— however 'nothingness' is always there, because it is never there. Moreover nothingness, or the infinite void, is always greater, bigger and larger than that which resides in the void, such as material particles. In this way there is a hierarchical relationship between Nothingness and Being, subordinating Being to Nothingness. Nothingness is eternal, infinite and all encompassing, while Being is temporary, finite and engulfed and surrounded by Nothingness. In a subtle and clever way therefore, atheist thinkers position nothingness as a reality that is superior to being. However, this is incorrect both from a scientific as well as a logical and philosophical point of view, absolute nothingness absolutely does not exist, which is of course the exact and literal meaning of this proposition. Nothingness, as employed by atheist thinkers, therefore refers to absolute nothingness, the total absence of anything or any type of existing reality. Theism, to the contrary—and in particular Vedanta theism—employs the term nothingness as a relative notion, and as a function and an aspect of being.

The second proposition follows logically from the first proposition, which states that if nothingness does not exist, then it cannot produce anything. The former cited Roman philosopher Lucretius (99-55 BC) stated: "Nothing comes from nothing." However, by employing the term nothingness in the absolute sense, nothingness is effectively turned into somethingness. This is exactly where the confusion starts. The singularity is a good example of this, for the singularity is nearly nothing, and is an ingenious way to create an intermediary step, whereby nothingness is turned into some form of somethingness. A somethingness that can actually play a part in reality. Of course, just based on our common sense, on any level it is absurd to think that everything, the entire universe, came from actual nothingness, from total and absolute non-existence. Even though our common sense can sometimes mislead us, in this instance that is not the case. While the proposition itself is already absurd, the logic of its refutation is inescapable, and reveals the absurdity.

The third proposition refers to the absurd logic that if nothingness did somehow exist, and if it did somehow have the ability to create something, there would still be no reason, logical, causal or otherwise as to why non-existence would transform itself into existence. The physicist professor Stephen Hawking, cited at the beginning of this chapter, formulated this as follows. He made this statement when expounding on the possibility of the long sought after and evasive unified field theory of physics:

"Even if there is only one possible unified theory, it is just a set of rules and equations. What is it that breathes fire into the equations and makes a universe for them to describe? The usual approach of science of constructing a mathematical model cannot answer the questions of why there should be a universe for the model to describe. Why does the universe go through all the bother of existing? Is the unified theory so compelling that it brings about its own existence?" [83]

The importance and role of 'nothingness' in atheism

Why is the 'existence' of nothingness so important to the atheistic worldview? The existence of nothingness basically makes atheism possible, for the following reasons.

Firstly, the existence of nothingness makes it possible to explain the origin of the singularity. By posing nothingness as the cause of the singularity, God can be excluded. As stated many scientists regard non-existence explicitly and implicitly as the origin and foundation of our universe, and they are of the opinion that the singularity originates from the void, from nothingness. Indeed, the very notion of the singularity evokes images of nothingness, emptiness, and non-existence. Effectively, the singularity is an almost non-existence, for it is an infinitely small point that exists within nothingness. The difference between infinitely small and nothingness is indeed infinitely small, next to nothing. In that sense the singularity is also a clever way to bridge the gap from nothingness to an entire universe full of stuff. By interjecting an intermediate step, from nothingness, to almost nothingness, to almost existence, to actual existence, to a subsequent explosion, to massive inflation ending up with an infinitely full universe, it removes some of the absurdity of the entire proposition. In the end, because singularity is not part of the universe (for at that point there is no universe), there are only two possibilities to explain the existence and origin of the singularity. The singularity either comes from nothing or it comes from something else, and that something else is God by definition (see the former discussion on this topic in chapter 5).

Secondly, nothingness creates—in the form of its close ally, empty space—the foundation for the pluralistic universe. Non-existence is associated here with infinite empty space that at a certain point is filled—at least partially filled—with energy and matter. However, this empty space will exist always apart from and next to the matter and energy that are residing within this empty space. Moreover due to its intrinsic infinite and eternal nature empty space is always larger and greater, both in space and time, then the matter and the energy residing in it. As mentioned before, matter and energy are sometimes there, and sometimes they are not there, while nothingness in the form of empty space, is always there because it is never there. The existence of this empty space is crucial for the materialistic vision. Without empty space particles cannot exist independent of one another, there can be no pluralism and therefore there cannot be chaos, coincidence and randomness. And these are exactly the ingredients that form the foundation of the materialistic world view.

Finally, there is a third reason why non-existence is so important for atheism and materialism. This has to do with the evolution principle referred to in previous chapters, which states that the origin of things must be simple, at least as simple as possible. The simplest of all things is non-existence. Based on this vision, emptiness—non-existence—would be a beautiful and very simple source from which all that exists, with all its complexities, comes from. Indeed, the process of evolution itself implies the creation of new 'things' and new 'attributes' out of nothing. The core of evolution, which we discussed in chapter 4 and 6, is that simple things gradually develop into complex things featuring completely new attributes. The only explanation for the origin and appearance of new attributes is that they come from nothing.

Everythingness versus nothingness

Based on these three arguments, it becomes clear how important the existence of nothingness is to the atheistic worldview. Atheism elevates nothingness to the highest principle of being and embraces nothingness as the ultimate explanation and the ultimate reality; nothingness, the infinite and eternal void in which a universe manifests itself, every now and again. This eternal and infinite void is always there; mass and energy are sometimes there and sometimes not.

Theism, particularly Vedic theism, on the other hand, rejects the notion of absolute nothingness, and instead replaces it with infinite fullness as the ultimate explanatory principle of the universe. God is the embodiment of infinite fullness, infinite being and the infinitely full reservoir of energy and attributes, both potential and actual. Instead of Supreme Nothingness, theism elevates Supreme Being to the highest principle of reality.

God is that Supreme Being and his 'completeness' and fullness is one of the definitions that is often used in theistic philosophies. In the Vedas, for example, God is called *Bhagavan*, He who is infinitely full of all attributes and properties. The conclusion is that the infinite universe came from something that is not nothing, but which is, in fact, infinitely full. After all, there is no other option. This is an interesting comparison of both worldviews that are both symmetrical, simple, and comprehensible. God represents the infinitely full, everythingness, atheism represents the infinitely empty, nothingness. The infinite fullness, theism versus the infinite nothingness, atheism: these are the two totally opposing options for the explanation of reality.

Something is also nothing?

It may sound incredible, but throughout the centuries there have been philosophers and thinkers who have disputed the existence of reality and of the world around us. These arguments vary from the thought that the world is only an illusion, that it is a projection of one's own consciousness, up to a fundamental doubt about anything. This doubt was taken away, at least to a certain extent, by the French philosopher René Descartes. Descartes formulated the historic proposition, a milestone in the evolution of human knowledge: 'Cogito ergo sum' or, in other words, "I think, therefore I am". Descartes argued that the existence of thought is the unquestionable proof of our existence. The idea was that doubt is also a form of thought and therefore doubting existence was a paradoxical confirmation of our existence.

Descartes, however, could not remove all doubt about existence. The Scottish philosopher David Hume suggested that even though our perception cannot be doubted, there is no certainty that our perception corresponds with the outside world. Hume argued that we do not directly observe objects in the outside world. What we see is a reflection of a reflection of an object. When I look at a bookcase, in reality I see an image of the bookcase in my head which I assume is a reliable reflection of the actual bookcase. In fact, this apparently simple observation is part of a very complex process with many intermediary transformations required. Firstly, I need light in order to be able to see at all. The function of light is that it is reflected by the bookcase, which absorbs the colours and shapes of the bookcase and reflects them in a coded form. The light is captured by the retina of my eye that accordingly translates

René Descartes

it to a new, unimaginably complex electrochemical code. This electrochemical code is sent to the brain, via the nervous system connected to the eye, where it is again translated. Somewhere in my brain there is then a conscious awareness of an image which we assume resembles the original bookcase. Therefore, what we actually observe is really an image of an image of an image.

Our other senses also work in a similar manner, and as a result, our consciousness is never directly in touch with the outside world. According to Hume this meant that our knowledge of the world is actually a representation of that world, technically called 'representationalism'. And as such, we cannot have 100% certainty that the representations are correct: whether they truly represent the objective world or even whether there is an objective world at all. Indeed, there is a school of thought that downright claims that the entire world is nothing but a mental projection, a construction of one's own consciousness. In philosophy this view is called 'solipsism', which literally means that 'only I exist'. In certain interpretations of the theory of quantum physics, this theory is also sometimes suggested.

Hume's position was considered to be rather extreme and was not considered realistic by many philosophers. His opponents argued that there are good reasons as to why we assume that our observations correspond with the world outside. First of all we are equipped with different types of senses that make simultaneous observations that support, supplement and do not conflict with one another. While I look at a bookcase I can also touch it and notice that the two different sensory observations correspond. Secondly, we are not alone and other observers are able to observe and describe the same bookcase, which fact helps to validate the existence of objects and the world in general. Thirdly the sensations I experience are imposed on me from outside. I do not have control over them, they are what they are. If the outside world was nothing more than a mental projection of my own consciousness, then I would be able to change the sensations that I am experiencing. If I didn't like the colour of the bookcase then I could give it a different colour, simply by wishing it. Unfortunately for all of us, this is not the case. The objective world is obstinate and stubborn whereby my senses merely observe what has been presented to them, without the ability to influence the content of what has been presented. It is a fact, therefore, that my observations of the objective world do indeed exist and that the conscious sensations that they generate cannot be doubted. The formerly cited British philosopher John Locke stated that there are certain sensations that cannot be doubted or disputed that are self-evident and that can only be explained from experience itself. He called them 'simple' ideas, such as the sensation of the colour blue, the taste of sweetness, or the sound of a piano note. These experiences have one thing in common: their existence is a self-evident matter-of-course and something we can only explain based on the experience itself.

When someone asks me what the colour blue is, then all I can say is that one cannot explain it, but only experience it. In other words, a simple observation is an observation or experience that cannot be divided into other experiences, cannot be reduced to other experiences or that cannot be explained by referring to other experiences.

Without placing too much focus on this, for there are more than enough books written about this topic, there are numerous indications that the objective world exists. Even the simple ideas that form the foundation of my empirical knowledge of the world outside, confirm that the objective world exists. The proposition that the existence of reality cannot be doubted could be called the certainty principle, supported by the principle of simple observations. Both principles confirm that 'something' is not nothing and that reality is not a hallucination; it does actually exist.

The Steady State: between something and nothing

There is however a theory that sits in between nothingness and being, the void and fullness and that does not need to choose between both ultimate realties. This is the Steady State Theory, which theory has been extensively discussed in previous chapters. This theory will be briefly revisited in this chapter, albeit from a different angle, focusing even more directly on the relationship between nothingness and being. According to the Steady State Theory, the universe is eternal in which matter and the void have an eternal co-existence. As a result, the difficult questions pertaining to the origin and the beginning of the universe have become irrelevant. But also the question as to why there is something rather than nothing, loses some of its relevance. Indeed, this is one of the big advantages of, and a key difference between the Steady State Theory and big bang cosmology: there is no moment of creation, or moment when nothingness has to transform itself into somethingness.

Furthermore, according to the Steady State Theory, the universe and the distribution of matter in the universe is uniform both in infinite space and the eternity of time. Both matter and space are eternal and infinite, however, in such a manner that infinite space is always larger than the infinite amount of matter that resides in this infinite space. This leads to another important implication of the Steady State Theory: empty space facilitates both the movement and independence of material particles with respect to each other as each material particle is separated from every other particle by empty space. This is the foundation of pluralism. The eternal existence of material particles for ever bouncing around in an eternal pluralistic universe, gives rise to the possibility—at least in theory—of creating unimaginably complex structures based on randomness and coincidence. Even the anthropic principle and the fine-tuned universe could have been explained based on the Steady State Theory, at least theoretically. According to the anthropic principle, which we devoted much time to earlier in the book, if the balanced relation between the four fundamental forces of the universe is not eternal, then it is statistically impossible that this balance emerged by chance at the first seconds of the Big Bang. But, if this balance is eternal and does not have an origin or cause, then statistical probability calculations no longer play a part. They just exist, eternally and without cause. Therefore, from the point of view of materialism, the Steady State Theory offers a suitable and symmetrical solution to a number of fundamental problems. For this reason atheist philosophers are fond of this theory and the big bang theory has been, literally and figuratively speaking, a big blow for atheism. Nonetheless, even in the scenario of an eternal fine-tuned universe, the second phase of the teleological argument, the emergence of complex living organisms, would still remain inexplicable and unresolved. Living organisms are non-entropic, they defy the law of entropy, not just once, but perpetually. While the constants that rule the universe, may account for the emergence of atoms, molecules, stars, planets and galaxies, yet they are fixed in their values, and fixed in relation to one another. The emergence of biological form cannot be explained on the bases of fixed constants. The inner workings of DNA in individual cells, as well as the interactions of numerous cells operating within a multi-cellular organism, and their reproductive capabilities are all beyond any mechanistic explanation. To quote mathematician Richard Thompson, the chance emergence of complex life forms within the 4.5 billion years the earth has been in existence, is $10^{150.000}$, not just a negligible number, but in fact an impossible odd.

Despite these statistical problems, the Steady State Theory does match our day-to-day experiences. We see that (apparent) empty space exists next to matter that is situated in this space. Matter is situated in empty space and it is situated in time; at least, that is what we perceive at first sight. We also observe that empty space enables material objects to move within this empty space. Our common sense tells us that something that is completely full does not leave room for movement. In order to drive on the highway the road has to be relatively free and empty of other traffic and obstacles in order to get to my destination. Space makes movement possible, or at least this is what we experience.

The Steady State leads to nothing

So, how is it possible then that a theory with so many elegant solutions and so much apparent symmetry nevertheless fails? The scientific reason is that the physical and cosmological data contradict the Steady State Theory in such a convincing manner that the theory cannot be sustained. The initial doubt came from Einstein, when he noticed a contradiction between general relativity and the steady state model. He attempted to solve that by introducing the "Cosmological Constant" theory which initially failed and was later revived, but in a different context. As was discussed in previous chapters, the observation of Edwin Hubble in 1927 of an expanding universe was a significant turning point, undermining the Steady State Theory. However, the definitive scientific nail in the coffin of the Steady State Theory was the discovery of cosmic background radiation in 1965 by Penzias and Wilson which confirmed the prediction that the big bang would have left a permanent radiation echo in the universe. Based on all the scientific evidence, the big bang theory emerged as the clear winner in this battle for the most accurate story of creation. At the same time, however, it has left exposed the deep flaws, from a logical, philosophical and scientific view, of an atheistic explanation for the creation of the universe.

However, the Steady State Theory also had other serious problems too, that have been discussed in previous chapters. In the end, the most important scientific and philosophical reason why the Steady State Theory fails is the fact that the theory assumes a pluralistic universe; a universe consisting of empty space, partially filled with matter and energy. As has been concluded in previous chapters, God does indeed exist, and his existence implies that reality is in fact a fundamental unity in which there exists no void. Without the existence of void, non-existence is not an option and the possibility of a pluralistic materialistic universe is eliminated.

Edwin Hubble

Back to square one: why is there something instead of nothing?

Having relegated the Steady State Theory to the dustbin of history, we again return to the choice between something and nothing; why is there something instead of nothing? This brings us to the third aspect of the all-versus-nothing discussion—why non-existence or nearly non-existence would transform itself into existence. This is a crucial question which we already partially addressed, amongst other topics, referring to the quote from Stephen Hawking.

In discussing this issue, we need to keep in mind two important points. First of all, we need to determine that such an event (the transformation from (nearly) non-existence to existence) is based purely on coincidence. It cannot be a necessity because the laws of nature do not yet exist and necessity can only be based on laws. Since there is not a necessity, existence cannot have a purpose. It only exists for itself, for its own existence, and the continuation of that existence.

Secondly, and basically all scientists agree on this, the universe will end one day. It is possible that the end of the universe can lead to a new big bang and the creation of a new universe—the big bang cycle. It can even be that there are an infinite number of parallel universes in completely different time-space dimensions that we will never get to know. In all cases, however, we deal with the same dilemma; non-existence that transforms itself into existence.

The principle of effectiveness or least action

It is helpful for us to look at this question from the point of view of effectiveness, because effectiveness plays a major role in the fundamental laws of nature. Stronger still, according to the mechanistic, materialistic worldview that developed after Newton, the pursuit of effectiveness and efficiency is an all-determining principle. This means that objects only become active when they are influenced by another force, therefore, out of sheer necessity. According to the first law of Newton—the law of inertia—an object that moves through empty space will move in a straight line, without changing its speed, unless a force influences the object and forces it to change its course or makes the object slow down or accelerate. In other words, an object will not make its course more complicated, by changing its direction or speed for no reason. In this respect, effectiveness can be defined as only doing what is absolutely necessary and unavoidable to achieve your goal with the least possible amount of effort. The law of entropy is quite similar in this respect, for according to entropy material particles will always find the least complex configuration in relation to each other, which configuration should end up in an equilibrium.

Reasoning along these lines non-existence is considerably more effective than existence. Non-existence does not require any effort, while existence requires enormous effort that is furthermore characterized by an unpreceded level of complexity. The universe changes constantly and violently and is in a permanent state of destructive restlessness. Since the big bang, the universe has experienced 13.7 billion years of cosmic explosions, enormous nuclear events, exploding supernovas, implosions, all-devouring black holes, and destructive radiation from quasars. In our own small solar system, we are very worried about our future from threats such as a possible cataclysmic meteor impact on Earth, which has happened before. Approximately 65 million years ago a gigantic meteorite hit Earth and started a nuclear winter leading to the extinction of many species of life at the time including the dinosaurs, —not that this is necessarily something we should be upset about. The dinosaurs were, at least according to most Hollywood film scripts, pretty obnoxious animals and, according to evolutionists, they paved the way for the evolution of man.

Indeed, the future of the universe is anything but rosy. Life on Earth will certainly end as soon as our sun has burned up all of its remaining nuclear fuel, in approximately 5 billion years. After first swelling-up, our sun will cool down and all life on earth will inevitably be destroyed. Earth will become a lifeless, ice-cold lump of matter, floating around in a universe that is

Ludwig Wittgenstein

itself gradually getting colder and colder. It is expected that, after many billions of years of further expansion, our universe will simply consist of ice-cold empty space with a piece of matter or some dismantled atoms floating around here and there. The 'Big Rip' scenario is currently the more prevalent theory about the future of the universe.

Therefore, all in all, existence is a useless effort that leads to nothing, at least it literally leads to almost nothing. Not that the universe should be bothered about this. The universe does not have any emotions and does not feel pain when the next destruction or nuclear explosion or icy cool-down occurs. These violent and disastrous conditions are described from the perspective of human consciousness and human emotions. Indeed, it will only be the conscious beings on Earth, and possibly other parts of the universe where conscious beings may reside, who will experience this development as painful. The universe itself will not be bothered by it. However, one can still ask the legitimate question as to why nothingness, the subsequent singularity or the pre-universe, would start creation. There is no reason for it, there is no necessity and it leads to nothing. Doing nothing and creating nothing would be more effective, or to put it in other words: if the universe were optimally efficient, it would not create itself. The German philosopher Leibniz (1645-1716) asked himself, 'Why is there something instead of nothing?' and an even more famous German philosopher of the 20th century, Wittgenstein, said in 'Tractatus Logico-Philosophicus':

> *"The mystical is not how the world is, but that it is."* [84]

Another example of this hunt for effectiveness is what philosophers and scientists call 'Ockham's razor'. This principle was conceived and formulated by the 14th-century philosopher and monk William of Ockham (1285-1349). It boils down to a preference for simplicity to explain complicated phenomena. When there are two explanations that offer an equally good solution, then the least complicated one should be preferred. In other words, we should not make matters unnecessarily complicated. Ockham's razor is a reflection of reality; reality will always seek the simplest solution. When we apply this principle to the matter of existence versus non-existence, then non-existence is by far the simpler solution.

Within the scope of the materialistic, mechanistic worldview the preference for effectiveness is also defined as the principle of least action which can be described as follows:

> **"The laws of nature are led by the urge for effectiveness, the principle of least action. This means that objects and forces only act based on sheer necessity, and that this occurs with a minimal amount of effort and with a maximum yield."**

This principle is based on the mechanical worldview and its logical consequences. In classical mechanics this principle is referred to as the principle of least action, and simply states that when an object moves from A to B, it will follow the path that requires least effort, the least amount of time and the shortest distance. Effectiveness, taken from the point of view of our human consciousness, of course coincides with achieving some sort of goal. In this regard, work, action, failure, problems and solving problems are only means to that end. However, if there is no goal, except for existence and survival, then the effectiveness of existence can be strongly doubted. Factually, non-existence would be more efficient. The conclusion according to this line of reasoning is that atheism, from a philosophical point of view, is not possible or at least very improbable. As on overall summary of this chapter and coming back to the propositions stated at the beginning of this chapter, nothingness does not exist, nothingness cannot produce anything and finally, even if it could produce anything, it would be completely illogical and inefficient if it were to do so. This implies that there is no way that the universe, in the end and after some intermediary steps, such as the singularity, came out of pure nothingness. It is simply absurd and impossible.

Chapter 9

WHY DOES REALITY EXIST IN THIS PARTICULAR MANNER?

"The fact that experiences occur in restricted centres and take the shape of limited, specific this-ness is eventually inexplicable."

F.H. Bradley, (1846–1924) British philosopher in
'Appearance and Reality' (1893)

The basic structure of reality

Let's assume, for arguments sake, that despite the arguments presented in the previous chapter, nothingness was still able to create something, in this instance a universe. If that is the case, why would nothingness have created this specific universe, with its specific attributes, exactly the way it is. Indeed a question that for centuries has mystified philosophers and scientists.

According to Wittgenstein, the fact that reality exists is a mystical fact. But why it exists in this specific and very complex way is just as mystical. The British monistic philosopher F.H. Bradley referred to this in his book 'Appearance and Reality' of 1893. Bradley stated that there is no explanation as to why experiences and objects take on a specific and limited form and why they have specific attributes. It is indeed inexplicable if these forms and attributes emerged out of nothingness, or out of something that is fundamentally devoid of form and attributes. Yet, reality in all of its complex diversity with its specific objects and specific attributes does exist, so there must be an explanation why it is the way it is.

To delve into this explanation we need to understand what reality is made of, and what its basic structures and components are that bring about its specificness and distinct attributes. In this context it is interesting to analyse the basic structures of reality and the universe from the perspective of Vedanta philosophy. The pre-eminent Vedanta philosopher Ramanuja (1017-1137) explained in his commentaries on the Vedanta-sutra, that reality consists of three essential components:

- Dravya in Sanskrit, (substance, energy)
- Guna (attributes)
- Karma (movement and change)

Dravya, substance or mass, is the foundation and basis of reality, which in the end consists of pure energy and consciousness. *Dravya* implies pure existence in its most basic and fundamental manner. Dravya expresses itself in time and space, for existence needs 'room' to exist, so to speak, and this 'room' manifests itself as time and dimensional space. Space and time represent the extension functions of existence: continuity, the continuous flow of time and contiguity, the existence of a contiguous substance within space. While Dravya implies pure existence, such existence is not devoid of attributes like shapes, colours, sounds, tastes and tactile attributes. Attributes are called Gunas in Sanskrit, and they qualify the underlying substance, creating distinction and specificity. Attributes distinguish one substance from another, because attributes by their nature are diverse and specific, expressing themselves in an infinite diversity of tastes, colours, sounds, shapes and odours. Substance forms the substratum of attributes, and while substance is one, attributes are many.

Karma means motion and change, and expresses itself in the motion of objects in space, as well as in the interaction of the multitudes of attributes, and the subsequent changes they bring about. For two chemical elements, hydrogen and oxygen for instance, to interact, they need to move and be joined together. Karma therefore also represents energy, needed to cause motion or bring about changes. According to Vedanta philosophy, change and motion is in the end, brought about by consciousness and intelligence, and leads to the organized complexity in the universe. Below is a diagram that displays these components, and their subsequent effects.

These components, substances, attributes and motion are perceived by us on the bases of elementary, empirical observations, or 'simple observations' as was explained in the previous chapter. This term was introduced by the brilliant British philosopher John Locke, in which he referred to the type of observations that are both elementary and which cannot be doubted. In the previous chapter some explanation was given on this issue in the context of the arguments of David Hume,

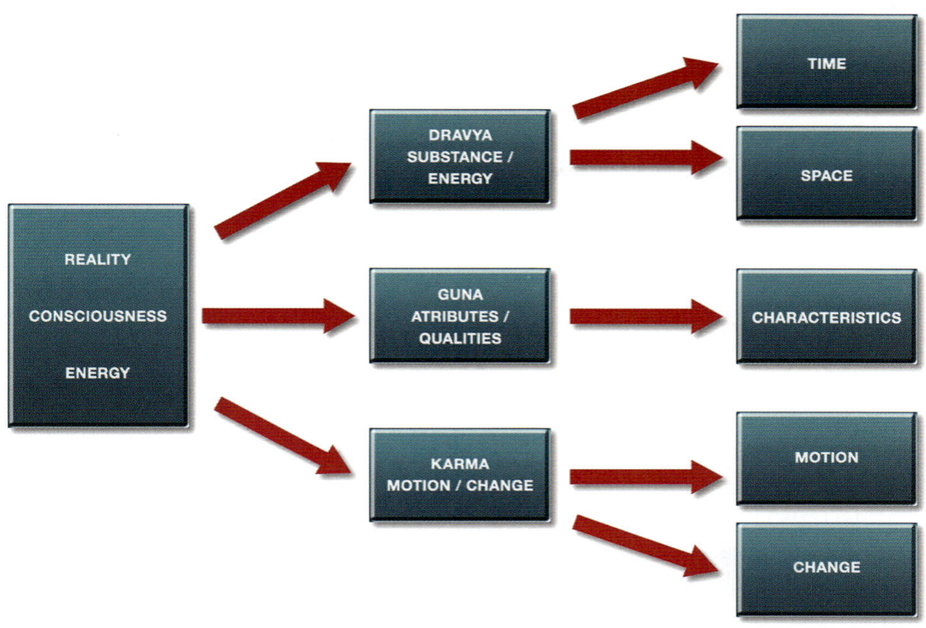

who had raised doubt about the validity of sensory experiences. In addition to the notion of simple observations, Locke also distinguished between primary and secondary qualities, where primary qualities relate to substance and motion, and secondary qualities relate to properties.

The key difference between primary and secondary properties is that primary properties are directly related to substance and they inhere in the substance. They are therefore considered to be objective, and provide objective information about an external object. Secondary properties such as sound, sight, taste, smell and touch are considered to be the result of the interaction with a conscious observer, and are not necessarily considered to be an inherent property of an object. While they may be caused by an external object, these properties do not necessarily inhere in an object, but rather they appear to be superimposed onto the object. The distinction between primary and secondary qualities, particularly those attributes created by light and sound, became even more pronounced in the 17th century. By that time it was discovered that light and sound are transmitters of information that represent objects, but are essentially different from them. It appeared that these transmitters add new properties to an object that do not necessarily inhere in the object itself. Colours are a good example of this, for colours are essentially no more than particular frequencies of light, which frequencies are interpreted by the eye, the brain and the conscious observer as a particular colour. This process seems to indicate that the colours we see in objects, are really superimposed onto these objects, but do not inhere in the object itself. What seems to inhere in an object is the chemistry of its surface, that, when showered with light, generates a certain frequency once that light has been reflected of its surface. In other words colours are more so the result of the properties of the transmitter of information, i.e. light, rather than the actual properties of the object itself. This division between primary and secondary qualities, is also reflected in the natural sciences, where, broadly speaking, physics deals with the primary qualities of matter, and chemistry deals with the secondary qualities.

In its effort to objectivize reality, science tends to focus on that which is objectively real, and can exist so to speak, independently of the conscious observer. Matter, energy, motion, time and space fit neatly into this category, for empirically and cognitively we sense that these realities exist independently of any observer. These were the primary qualities referred to by Locke. The attempt to objectivize reality becomes complicated however when confronted with secondary properties, such as colours, sounds, odours, tastes, heat and cold, and touch. It introduces the role of the conscious observer not as merely passive and objective, but as an active participant, that may even alter or add properties to the object observed. The reduction of properties to a subjective experience only, generated by consciousness, becomes even more relevant since this notion seems to be supported by the inner workings of chemistry, the domain of secondary properties. The periodic table of elements seems to have reduced the array of chemical properties, such as colours, smells, tastes or sounds to the quantitative numbers, arrangement and configuration

278

of sub-atomic particles. Particles that are apparently devoid of any of these properties, but somehow are able to manifest them when the right number and configuration is there. The periodic table of chemical elements lays that out very neatly. This fact further enhances the notion that reality is indeed composed of an objective part, particles possessing mass, shape, motion and location, and a subjective part, subjective interpretations of properties that are generated by these particles, but that are yet and at the same time quite distinct from them.

Some philosophers, such as the British philosopher and mathematician Alfred North Whitehead (1861 – 1947), have labelled the distinction between primary and secondary qualities the bifurcation of reality into that which is objectively real and that which is subjectively real. He was critical of this bifurcation, for reasons that are in some ways similar to what is being discussed here. While it is true that consciousness plays an augmented role in the perception of secondary qualities, consciousness plays a role in every form of perception, including the perception of primary qualities. The key point is that properties, whether primary or secondary, are never perceived separately from their underlying substance, and are always observed in connection with its underlying object. Therefore all properties, both primary and secondary, do inhere in the object, and in the end they are simultaneously both one with and distinct from its underlying substance. This relationship of simultaneous oneness and difference, in this instance between substance and its properties, has been discussed in previous chapters and will be extensively explored in Chapter 12.

It has also been argued by some scientists and philosophers that all attributes, both primary and secondary are real only, due to them being observed by a conscious observer. This point has been discussed already in the context of the Copenhagen interpretation of quantum physics. In this interpretation the mistake is made that reality is in fact considered to be the product of an individual conscious observer. That is not correct, but it does highlight the great mystery of the interaction between consciousness and the outside world, the objects of consciousness.

A mystery that will be further explored in the next chapter. While consciousness is essential in making an unknown existence known, this does not imply that consciousness is itself the creator of that existence. Our experience of the world as a conscious observer is presented to us, and we have little or no control over that which is presented to us, and how it is presented to us. In other words, there exists an objective reality that interacts with a conscious observer in different ways and in varying degrees, but yet exists independently of that observer.

Let's now start our little thought experiment, again making use of the bookcase as our object of observation, as per our experiment in the previous chapter. In doing so our observations can be divided into a number of simple categories. These are the categories as indicated in the scheme further below, and as mentioned before, go beyond the categories and constituent components of physics and chemistry. Starting with our consciousness, we observe our own existence and the existence of an objective outside world. Accordingly, we observe that reality is diverse and consists of several fundamental parts. Looking at the bookcase, we observe an object that exists, and is made of a certain substance—in this case wood. Furthermore, we observe that the bookcase possesses specific attributes, such as shape, a colour, a smell, a certain roughness, and a certain temperature. The bookcase does not move at this moment, but conceivably it could be physically moved or more drastically it could be burned to ashes. In either case weather moving it or burning it, the bookcase will be subject to all sorts of changes. Within the atomic structure of the bookcase there is a massive amount of energy stored in accordance with the formula $E=mc^2$. We also observe that the bookcase takes up space, and that it continues to exist for a certain period of time. Finally, we notice that the bookcase possesses a level of complexity, in this instance designed by a furniture designer. Coincidentally a repair man comes in who is about to replace a broken lock. The repairman, just like me, is a conscious observer, part of the objective reality. In summary, these are the components that constitute the bookcase, in this instance as a representative of objective reality:

- Energy in the form of mass, the substance that reality is made of
- Energy in the form of forces, the energy that causes change and movement
- Attributes such as smells, colours, shapes and sounds
- Movement and Change, the process of change itself
- Space
- Time
- Organised complexity, the interaction between all above mentioned components
- Consciousness, the conscious observer and organizer

As mentioned before, this analysis corresponds to some of the basic tenets of Vedanta philosophy, and can be further summarized in the following diagram:

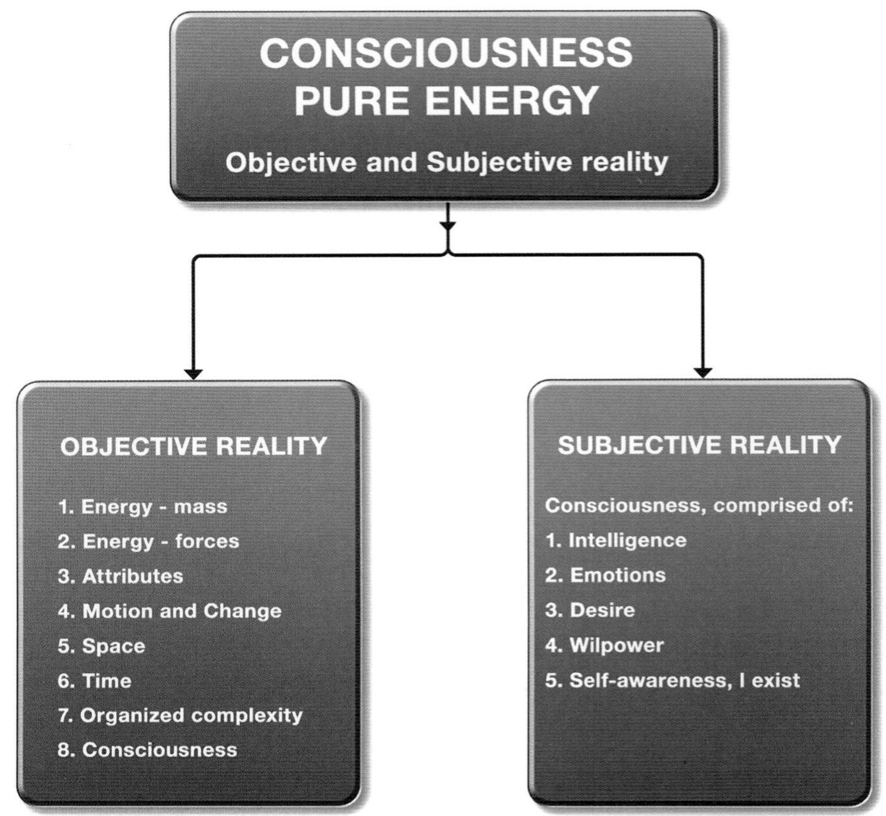

All these components refer to the objective reality as we observe it with our consciousness. In this way reality is dissected in eight categories, which are all unique and fundamental. This means that they are part of the basic structure of reality and that they define the nature of reality. We cannot explain these complexities in a reductionist manner, for they have no cause. We can, however, determine their existence, their effects on reality and their mutual interactions.

Our consciousness also reveals an inner subjective reality. This subjective reality manifests itself in various ways and at various levels, such as intelligence, emotions, desires and willpower. From our consciousness we confirm our identity, our realisation that we are a person, and also the objective existence of the world outside. The diagram above reflects these different, fundamental aspects of existence. They too are fundamental in that that there is no causal explanation for their existence.

These components and functions of existence generate a reality that is very complex, and it appears that existence itself, by its very nature, is inherently complex. Atheist philosophers have no explanation for the existence of these complexities, other than that something 'simple', such as nothingness, transformed itself inexplicably and illogically into something. A something which, from all the infinite possibilities, turns out to be hugely complex. The how and why is not only illogical, but it is furthermore totally inefficient. Apart from all other arguments, for this reason alone atheism is very improbable.

Why is there not a simple and undifferentiated existence?

Continuing our discussion above, and looking at it from an from the atheistic viewpoint, if existence exists, then why not in the simplest way, just pure matter, substance or energy without further attributes, structures and dynamic changes? This again would be the logical conclusion from the point of view of effectiveness. This form of existence could be compared to the singularity, the form of existence prior to the big bang, in which existence—at least theoretically—is a mere condition of indifferent, pure energy. The fact is, however, that the singularity is not the very simple form of existence that it theoretically was intended to be. In chapter 6 we discussed this thoroughly and substantiated that the singularity by default cannot be a simple form of existence. When the singularity subsequently changes into a big bang from which the universe originates, simplicity is nowhere to be found and is gone forever. The universe as we know it, is dynamic and infinitely complex.

But even if the singularity, again for arguments sake, were an undifferentiated, simple form of existence, then there is still no causal reason why this pure energetic form of existence would change into a complicated and dynamic universe. Not only is it illogical and devoid of any causal necessity, moreover, it is highly inefficient in the grander scheme of things.

All our scientific knowledge confirms that the universe is very complex. Moreover, as our knowledge of the universe increases, so our realisation of this complexity grows exponentially. Reality is fundamentally complex right down to the deepest layers and roots of its existence. Compared to the atheistic explanation for the existence of reality in this specific and complex form, the theistic explanation refers to the infinite 'fullness' and complexity of God. While God is one, within this unity there exists an infinite diversity of energies and an unpreceded level of complexities. This fullness and complexity is the nature of God and it is reflected in his energies. The 'fullness' and complexity of God is the pattern on which each part of his creation is founded. The structure of his creation and his energies is in many ways similar to a holographic projection. One of the characteristics of a hologram is that each part, each building block carries the totality of the hologram in it. The complexity and nature of the whole are reflected in each part which is an appropriate comparison.

In the next paragraphs these various levels and types of complexities are discussed and dissected. Key realization is that existence itself, just in order to exist, on the most fundamental level, is complex. These fundamental complexities are irreducible, they just are and therefore cannot be explained any further. It is for that reason that Ramanuja's analyses, breaking down reality into its most fundamental components, substance, attributes and motion, is so revealing. For no matter how simple these components are, they reveal simultaneously that existence by its very nature is complex, which complexities are moulded by the essence of reality itself.

Different complexities

The first complexity – energy and mass

The first complexity refers to the fundamental building blocks of existence, or what we call mass, substance or material. It is what gives reality substance and it is the carrier of all other attributes. Since mass is the same as energy—and vice versa—this first complexity is therefore defined as energy and mass. Energy, in this instance, refers to the 'Being" function of reality. We experience mass and energy daily, our world is made of it, we are made of it and the universe is made of it. The complexity of the existence of energy and mass is directly related to the question discussed in the previous chapter; why is there something instead of nothing? The very fact that energy and mass exist, inherently creates the first complexity of reality.

The second complexity – energy and forces

This second complexity refers to energy, in this instance referring to its function of force and motion. Energy, in the sense of "force", is something that exists, both inside and outside, of mass and energy. We have already discussed the four fundamental forces of the universe in depth, which are gravity, electromagnetism, the strong nuclear force and the weak nuclear force. Gravity is a force in the form of an attraction between two physical objects containing mass. Electromagnetism is the force that binds electrons, the particles that spin around the nucleus of the atom and give the atom substance. The strong nuclear force binds the particles of the atomic nucleus. The weak nuclear force facilitates a number of crucial processes that occur in the nucleus, one of which is the process of radioactive decay.

We all constantly experience gravity, for example, when we drop something or because of our own body weight. We can also experience electromagnetic force, for example light and electricity. We can only experience strong and weak nuclear forces in an indirect manner, since they occur on a subatomic level at a very small scale indeed. A force is a field of energy whose presence is everywhere within a certain space and which has the characteristic of contiguity. Based on the mechanisms of Newton, the Laws of Motion, we can understand how matter influences other matter, but gravity, electromagnetism, and nuclear interactions are far more difficult to explain, for they seem to be devoid of material substance. How does one object gravitate to the other object, while there is only empty space between the two? Einstein defined gravity as a 'curvature of space', but how can empty space have a curvature? In the end, scientists have concluded that space is not empty but filled with energy. The existence of forces and force fields have made existence infinitely more complex, and yet existence without them would be inconceivable.

The third complexity – attributes and characteristics

The third complexity represents the existence of attributes, for example colours, sounds, smells, taste, etc. This is how material particles are distinguished from one another and are qualified on other grounds than where they are (temporarily) located in space. Material objects are identified by and have their unique character because of attributes which distinguish one object from another.

It has already been discussed that the observation of attributes, for example the colour blue as an attribute of my blue shirt, is mostly subjective and largely produced by a conscious observer. But the fact that my shirt appears blue is due to the object itself – the shirt – combined with the light waves that are reflected by the shirt which transfer the image of the shirt and the colour. The colour 'blue' is produced by a certain frequency of light waves that are translated by my brain and assume the characteristic of 'blue' within my consciousness. But why does a certain frequency generate the perception of blue in my mind and, even more so, what is blue? We know that it exists, for we experience it and we cannot doubt this experience. It is what it is, it is the way it is, and on a most fundamental level, the colour blue, or properties in general, cannot be reduced to something else; they are irreducible. Therefore there is no logical or causal explanation for their existence and their origin. Attributes increase the level of complexity significantly, for attributes by their nature are diverse, specific and multifarious. The universe is full of properties, expressed in an infinite array of colours, sounds, tastes, shapes, odours, and displayed in an infinity of individual and specific objects.

The fourth complexity – movement and change

The fourth complexity relates to movement and change. Movement is the consequence of forces that were described under complexity two. Forces lead to movement of matter and they lead to changes in that movement. Change is also linked to the third complexity, the existence of attributes that are constantly changing and interacting. This happens, for example, when two chemical substances mix and interact, creating a new substance with totally different attributes. Matter, energy, and attributes change perpetually, and they influence each other constantly. Because of a better understanding of the laws of chemistry and physics, we are able to predict the outcome of chemical interactions. Yet, predicting the outcome of a chemical reaction, does not mean we understand the why of that outcome in a fundamental sense.

An important aspect of defining change is the energy that is necessary for making change possible. Where does this energy come from and why does existence contain motion and change? In the next chapter we will discuss the creative force in the universe, namely consciousness. Consciousness has several functions, amongst which is willpower. Willpower is a force that is generated from within and does not have a cause; in other words, it is 'The Unmoved Mover', that which causes movement but does not move itself. There is only one force, one phenomenon in the universe that has this characteristic and that is consciousness. This faculty of consciousness is also an example of the *Acintya* principle, since consciousness moves and simultaneously does not move, it changes and simultaneously does not change. The *Acintya* principle, as explained before, refers to the fact that God in his Nature and Being contains contradictory attributes. The most striking paradoxical characteristic is that whilst he is one with his energies and creation, he is simultaneously different from his energies and his creation. This paradox is the foundation of energy, consciousness and personality, and in the end is the foundation of God.

The fifth complexity – space

In previous chapters we have extensively discussed the concept of space. We experience space as self-evident and matter-of-course, inextricably linked to existence and reality. Upon closer examination however, space is not an ordinary, self-evident fact and it is not what it looks like at first sight. We mostly think of space as the infinite, empty nothingness that is filled with energy and matter. This space is always there, eternally, absolute and undisturbed and forms the cosmic stage on which reality plays itself out. However, modern scientific insights provide a totally new understanding of space. Einstein defined space as a relative reality, closely linked to and dependent on time. He called this the space-time continuum. More recent discoveries stated that space is not empty at all, but filled with all sorts of energies, varying from gravity to dark energy. This energy is also called vacuum energy to highlight that space, on a most fundamental level, is not empty and is therefore not vacuous but filled with energy. Even the word 'filled' is not completely correct, since space is regarded as a function and extension of energy. At its core, space is a function of existence, of being; it is the extension function of being.

The sixth complexity – time

Time is, like space, one of the cornerstones of the cosmic construction. We also regard time as a matter-of-course and, just like space, we consider it to be eternal, absolute, and infinite. However, time is also not what it seems at first sight. Einstein considered time to be a relative concept and linked it to space. The common perception is that time consists of the past, the present and the future and that time moves irreversibly from the past to the future. Present is the apparent wafer-thin point where the past and the future come together. This is incorrect however. The past and the future only take place in the present and are part of the continuous present. This does not imply, as some scientist and philosophers claim that time does not exist. Time does exist, but in the shape of the eternal present. Time is, just like space, a function of existence and time is, just like space, the extension of the function of being.

The seventh complexity – the anthropic principle: unimaginably complex structures

In chapter 2, "Chaos or intelligence – the teleological argument", we discussed the complexity of the universe thoroughly. We discussed subatomic particles up to mega star systems. The presence of the first six complexities described here, such as energy, mass, attributes, forces, movement, space and time have created a universe of infinite diversity. Diversity in its turn can either lead to chaos or organised structure. The teleological and anthropic principles confirm that the diversity in the universe is extremely organised and balanced. This implies that within the diversity and multitude there is a coordinated, central, and intelligent unity. This complexity can only be explained based on the existence of a central and intelligent coordination of the universe. Even though organized complexity is in itself a combination of other fundamental components, such as mass, energy properties, change and intelligence, it is such a distinct feature of reality that it has been promoted into a category of its own.

THE ULTIMATE COMPLEXITY – CONSCIOUSNESS

"I could imagine that I—in a delusional state— would deny the existence of the physical world outside. But I cannot imagine at all how I could deny the existence of my own consciousness. It simply is. I therefore consider it as totally fundamentally linked to my body, however I consider it to be even more fundamental than my body. Knowledge of consciousness is elementary, just as inalienable from my existence as water from the ocean." [85]

Bernard Haisch in 'The God Theory'

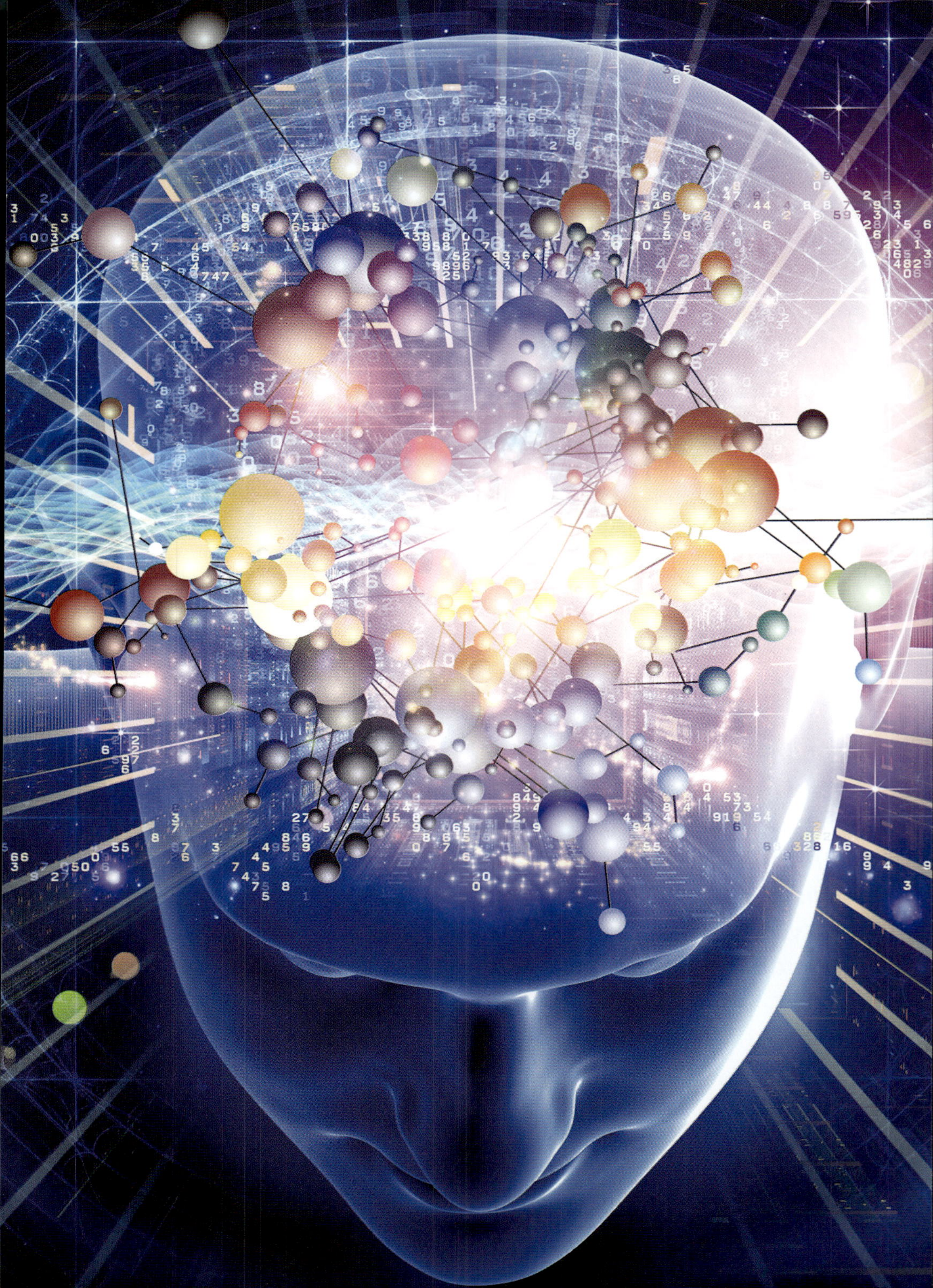

The importance of consciousness

While writing this book I am aware of the fact that I am conscious, and that because of the presence of consciousness I am able to understand, analyse and communicate what this book is about. I write this with the knowledge that the persons reading this book are able to understand my words, sentences, and concepts with their own consciousness. This book is only meaningful in every sense of the word because of the existence of the consciousness of the author and the existence of a comparable consciousness within the reader. Similarly, research and the development of theories into the origin of the universe and the meaning of life can only occur when there is a presence of consciousness. Our very existence is defined by the fact that we are conscious beings. The importance of consciousness can therefore hardly be overstated and overrated. Despite its importance, consciousness is at once also the most mysterious of all mysteries, the most intriguing of all intriguing things and the most complex of all complex phenomenon. It is also, strangely enough, the most self-evident, the most obvious and the simplest phenomenon. For it points to a single and indivisible unity: my existence as an individual, where all complexity and diversity converge by means of my consciousness into an indissoluble unity. That unity is the individual, the Self, the I: an indivisible reality. We experience ourselves unmistakeably as an indivisible individual with a unique identity. I experience that indivisible identity at this very moment, but it also extends beyond the present into the past. When I look at my life, from my first childhood memories until the present as an adult, my life, my body, my mind, my circumstances have all changed significantly in every way. There is, however, one thing that does not change, regardless of the changes and that is the unmistakable sense and feeling of 'I', my identity. My consciousness, my memories, my emotions but also my body have all undergone major changes. What does not change is the I, the person that has observed these changes and who is aware of all the phases in those changes; this I remains fundamentally the same. On the other hand, consciousness is extremely complex. It is able to process millions of impulses, impressions and data into something meaningful in a split second, and turn that complex process into what we call "understanding" or "comprehension". According to the Britannica the senses of the human body, such as eyes, ears, skin, taste and smell organs send 11 million bits per second of information to the brain for processing. From these 11 million bits only 50 bits are actually processed by the conscious mind, which means that the vast majority of information is filtered out by the brain and the sub-conscious mind. This massive filtering is accomplished by the 100 billion brain cells in conjunction with the parallel, synthetic processing capacity of consciousness.

Comprehension and computers

'Comprehension' is a core function of consciousness and it is an utterly unique and mysterious property. This property expresses the crucial difference between what we call intelligent life and inanimate objects. It also expresses the crucial difference between what has become a topical subject, namely the comparison between computer generated artificial intelligence and human consciousness. The current generations of super-powerful computers can easily beat man in terms of data processing speed. However, data processing has little or nothing to do with comprehending the data, or comprehending the results of the processed data. Data processing is entirely determined by the instructions of the programmes that are stored in the software of the computer. This programming is the result of the conscious act of a computer programmer who has created the computer software programme based on his or her understanding and insight. A computer does not 'understand' these instructions but follows the instructions of the computer programme faithfully and blindly. One of the consequences of this lacking of 'comprehension' is the fact that computer instructions are always processed serially; the next processing step can only take place once the result of the previous step is known. This is just the way it is; even powerful computers that work with parallel processers where several processing functions take place simultaneously follow this procedure. Each processor performs its instructions serially, in accordance with the instructions of the computer software. At certain points in the processing the various processors will exchange data with each other. However, this exchange also occurs serially, in accordance with the scheme that was established previously in the software. The crucial point about serial processing is that each processing procedure leads to a subsequent processing procedure that was determined in advance. A computer cannot deviate from the path of instructions that was determined in advance and a computer cannot, based on 'comprehension', make a different choice. Nevertheless, the enormous complexity of modern software combined with the power and speed of advanced processors means that computers can produce incredible results. Therefore, computers may appear to be intelligent, and may appear to have understanding and consciousness. However, this appearance of intelligence is only a virtual reflection of real intelligence, namely the intelligence of the human programmer. In the end, it is and remains a computer and it is therefore no more than an extension of human intelligence and consciousness.

Science fiction writers and filmmakers have written numerous scripts about powerful computers that take over the world. Serious scientists have also predicted a future of unimaginably intelligent machines. In these (often wild) stories, the super computer is portrayed as a devious and dangerous competitor to man. The super computer shows, almost without exception, Darwinian traits and is driven by self-preservation, the struggle to survive, and in the end the desire to be the strongest. Of course, it then regards man as the enemy and accordingly starts a battle where the weaker *Homo sapiens* loses. While fighting the super computer villain, I have always wondered why none of the embattled hero's came up with the simple, yet brilliant idea of pulling the plug out of the diabolic machine.

Contrary to computers, human consciousness is characterised by a real parallel processing of information, and is synthetic in nature. The synthetic nature of consciousness is expressed through the ability of simultaneous observation; the convergence and the coming together in space and time of large quantities of data. This simultaneous processing ability does not demand that all prior observations have already been processed and been brought to a conclusion. The result of information processing in humans, and the comprehension of it, manifests itself directly, immediately and simultaneously. Based on comprehending the content and conclusion of a process, but also on the bases of our will power, desires, and emotions, a conscious person can make any choice from countless possibilities, including a not pre-determined choice. This choice can be a rational choice, or several rational choices, based on logic, but people are also able to make irrational choices, as we all know. Besides the comprehension function, consciousness is also characterized by the presence of emotions, desires and willpower. These characteristics are evidently not present in a computer, something that even the most fanatic advocates of the unlimited possibilities of artificial intelligence will acknowledge.

Another interesting point of comparison is that while emotions do not necessarily contribute to logical solutions or speedy data processing, they do play a crucial role in the everyday life of people. In social, societal, and inter-personal matters especially, the emotional dimension plays a crucial part in understanding and solving problems; for a computer this is simply impossible. In essence, the difference between parallel and serial data processing is determined by the presence or absence of comprehension. In serial data processing each result leads to one possible follow-up step in the processing procedure. Crucially, that next step is pre-determined, having been anchored in the software such that the software instructions are followed blindly. On the other hand, the presence of consciousness, comprehension, emotion and will power changes everything. Each and every subsequent step in data processing that occurs with comprehension and consciousness is liable to change at any moment due to new insights or an expression of will power. Therefore, each result is unpredictable.

This remarkable ability of the simultaneous processing of information is possible because consciousness has the unique ability to 'broaden and expand' the point in space and the moment in time. The expansion of time has to do with the fact that we tend to misinterpret the concept of time. Often, the present, the now, is considered to be an ultra thinly sliced moment at the intersection of where the past and future meet. In reality, time functions in an opposite manner. Our memory of the past takes place in the present, just like our anticipation of the future takes place in the present. Past and future do not exist, except in our consciousness, as a memory or an anticipation. All our conscious observations and experiences occur in the present, the now. The question is how broad and deep is the moment we call 'now'. While writing this book, for example, I make use of information that I have learned in the past, such as my language skills. From my memory this information is transferred to the present immediately and simultaneously, where it merges with my consciousness of the keyboard that I type on and the computer screen that displays what I am writing. In this way, millions of images, neural activities, memories, and experiences come together at a specific time and place, namely the moment and place that I am writing right now. The moment in which all these memories and experiences come together is always the present, the now.

What is 'now' then, how long, broad and deep is the moment in time we call 'Now', and what is the meaning of the ability to 'broaden and expand' consciousness. While we experience and measure time in moments, these moments are not ring-fenced. There is no clear point or clear cut-off point where one moment ends and the next moment begins. The reason for this is that time is a continuum, as we discussed in the paradoxes of Zeno. As consciousness expands and develops, it actually broadens in time. Therefore, the present becomes broader and larger at the expense of the past and the future. The past and the future are basically swallowed up by the present and they become increasingly part of the present. In the end, the present becomes the all-embracing eternity, in which time is nothing but the experience of the eternal now. This also applies to our experience of space: while time is divided into moments, space is divided into locations. We are presently at a certain location and we experience that location. But again, one location is not clearly separated from another location. Space, like time, is a continuum where one location blends and merges into another location. My consciousness, as I sit here behind my laptop, expands over a relatively broad environment, my study room, my phone, the street I live on and the thousands of objects in my study room. I observe them all basically at the same time, one object perhaps clearer than the next. However, my realisation of the totality of the space that I am in is constantly present.

In comparison, my consciousness is far greater and broader than, for example, that of an insect residing in the seams of the wooden floor of my study. The more one's consciousness is expanded, a larger portion of the environment can be observed. Just like I can observe the entire room, while the observation of the insect is limited to a crack in the floor. In a similar manner, there may exist a being whose consciousness is far greater and more developed then mine, and that can observe the entire town I live in, within one single experience. The expansion of consciousness in space and time does not have any natural boundaries. The more developed the consciousness is, the larger the area and the moment of observation. In a way it is not easy to describe and visualize these properties of consciousness, especially because consciousness is so unique, incomparable and self-evident. All of the above properties, clearly distinguish consciousness from artificial intelligence in a fundamental way.

Consciousness is self-evident and obvious

The fact that consciousness exists goes without saying, yet we can only know consciousness because it reveals itself to us. We cannot escape our consciousness and look at it from a distance. Consciousness is not an observable object that we can compare to other objects. Much of our knowledge of objects comes from the possibility of being able to compare them to other objects. When we observe a cow, we know that it isn't a goat for they have different characteristics. On the other hand, they are comparable, though, for they are both members of the category of animals that have a number of common characteristics. Consciousness, however, cannot be compared to anything: what we know of consciousness is something we only know because of consciousness itself, without a comparison to other objects. In the category 'consciousness' there is only one member: consciousness. It is absolutely unique and stands apart from everything, and therefore it is incomparable. It is obvious, unmistakable and self-evident, since these words refer to something that reveals itself. Consciousness is exactly that: it reveals itself and it also reveals other things including our own consciousness. Consciousness is often compared to light, for light is self-evident, its presence is obvious and it makes everything else visible. Max Velmans, an English psychology professor, defined consciousness as follows:

"Everything we are aware of at a given moment, forms part of our consciousness, making conscious experience at once the most familiar and the most mysterious aspects of our lives." [86]

Consciousness is the creator of meaning

Everything we do with regard to consciousness and conscious observation, we undertake from within our consciousness. Even when we try to analyse our own consciousness, we do this with the help of that very same consciousness. All questions referring to existence, its cause, the meaning of existence, the matter of design versus chaos, suffering, and trying to make sense of life all arise from our consciousness. Consciousness is the ultimate creator of meaning within our reality. Consciousness is that which makes existence meaningful; indeed, it is the only thing that makes existence meaningful. Without experience and without consciousness, existence is literally meaningless, and it might just as well not exist. The search for knowledge, doing scientific research, the philosophical thoughts about happiness, suffering, meaning and design all arise from the same source— our consciousness. This search for knowledge is not executed by electrons, photons, the space-time continuum or by nuclear fusion, but by creatures possessed with the most important attribute that distinguishes and defines us: consciousness. Consciousness is the ultimate complication that, because of its extraordinary and incomparable attributes, is elevated above all else. It is completely unique and incomparable to any other phenomenon or reality.

Consciousness is the foundation of our experience of reality and it is the principle that makes us sure of our own existence and it makes us sure of existence in general. Descartes formulated his previously cited and famous statement: "I think, therefore I am". What he actually meant is: "I am conscious, and therefore I am." That is to say that thinking is just one of the functions of consciousness, as will be described in detail in the following paragraphs. Consciousness also comprises feeling, willing, desire, and the realisation of identity. Consciousness is not only the foundation from which we derive our existence, but also the certainty or sanctuary of our own existence and of reality. Because of consciousness we are able to formulate the certainty principle and the principle of primary and indisputable observations.

We are conscious, and we are situated within our own consciousness, but we do not have a clue what this means at a basic level. While we think about this, we also think about what our consciousness shows us: the world that exists. This existence is there, both externally (as observer) as well as internally; we are part of it and our role is not just that of observer anymore, but also of being an actor and participant of existence.

Functions of consciousness

When observing our own consciousness, we can ascertain that consciousness functions in different ways, and at different levels of clarity and intensity. These levels are the state of wakefulness; the sleeping or dreaming state; and the state of deep, dreamless sleep. These levels are characterised by the reach and intensity of consciousness in each state. In the states of deep sleep and dreaming, our consciousness limits itself to mainly internal mental processes and external sensory perceptions are put 'on hold'. In the state of wakefulness, our senses our switched-on, our mental processes are connected to our senses and to the external world. As we all know and experience, there is an enormous difference between being awake, dreaming while sleeping or being in deep sleep. In the dreaming state, for example, our perception of time changes dramatically; events seem to take place in a period of hours or even longer, while in reality the dream only lasts several seconds. This indicates the enormous diversity of conscious experiences and the large differences between the states of consciousness. Typically, when we dream, we experience the dream as reality, just as real as the state of wakefulness. The reason for this is that we do not know that we are dreaming, so we assume that the dream is reality. In a comparable way, whilst we are reading this book our current consciousness is in a state of wakefulness, but this state is also a type of dream in relation to a condition of expanded consciousness. The problem is, and this is the nature of consciousness, that we experience each level of consciousness as absolute and real. We often need others to focus our attention to our state of consciousness, for we tend to be 'blind' to our own state of consciousness. For that reason in Eastern spiritual traditions the help of teachers and gurus is recommended to help one to identify one's 'blind spot' and to help one progress on the spiritual path towards a higher state of consciousness.

Consciousness can also be analysed and described according to various functions and attributes, the most important of which are the passive and active functions. The passive function of consciousness refers to the process of observation and having sensations and experiences. This process can be subdivided into intelligent observation on one hand, and emotional observation on the other hand. The active function can also be subdivided into two separate functions, which are will power and desire. In addition, consciousness has an even more unique function, namely the fact that it is conscious of itself—or, in other words, self-awareness.

This analysis leads to the following categories and sub-categories of consciousness:

1. **Passive consciousness, which is sub-divided into:**
    a. intelligent consciousness, or thoughts
    b. emotional consciousness, or feelings

2. **Active consciousness, which is sub-divided into:**
    c. desires, or active consciousness that is driven by emotions
    d. will power, or active consciousness that is driven by both emotions and intelligence

3. **Self-awareness, or consciousness that is able to observe itself and that is aware of its own behaviour and consciousness**

Consciousness therefore possesses three attributes or functions that are fairly easy to recognize. With a little introspection, we can acknowledge and distinguish these functions within ourselves.

Intelligence

Because of the intelligence function of consciousness, we have experiences that are meaningful to us. This does not mean that we understand each experience, but it means that the experience appeals to our ability to understand something. Intelligence is the foundation of what we call the ability to 'understand' or 'comprehend'. Intelligence is the instrument with which we examine and analyse reality; it is the instrument with which we conduct scientific research and the instrument with which we develop philosophical theories. Without conscious intelligence there would be no science or philosophy and no ability to think, analyse, and remember, nor to arrive at a synthesis and conclusion. Intelligence is the driver of our thirst for knowledge and our curiosity to know why we exist, why existence exists, and whether or not God exists.

Intelligence is also the characteristic through which we can recognize and acknowledge the existence of God. The core of the teleological argument is the all-pervasive presence of intelligence in the universe. The complexity of the universe, the different levels of complexity as described in the previous chapter, and the complexity of interactions of the various energies in the universe are all manifestations of the underlying and infinite intelligence that drives and governs the universe. This infinite intelligence is a function of the consciousness of God in the same way that our intelligence is a function of our own consciousness.

Emotions

Our emotions make us aware of all the different emotive flavours that can be divided—for the larger part—into different degrees and shades of happiness and sorrow, joy and pain. Feelings matter; they touch our souls directly and they determine our psyche in a fundamental way. Our feelings also express our preferences and our dislikes. We long for pleasant, nice and happy sensations and emotions. On the other hand, we will do whatever it takes to avoid painful sensations and emotions. It is precisely because of this strict distinction and our clear preference for pleasure and happiness that we wonder why we are being forced to suffer (which we discussed in detail in chapter 7). Intelligent observation gives us objective information on the objective world and its objects, whereas emotions are subjective and reveal our internal state of being. While emotions can be caused by the experience of external objects, they do not describe these objects. Instead they describe our internal, emotional reaction to these external objects.

Desire

Apart from its passive functions, consciousness is also active. This is expressed through the attributes of will power and desire. Desire is mainly driven by emotion and it has an all-determining influence on the awareness of an individual. Indeed, the desire for joy and happiness and the dislike of pain and suffering are the two most determining desires for each and every person. Psychologists and philosophers agree in general that human consciousness and actions are essentially determined by two basic desires. In the 4th century BC, the Greek philosopher Aristotle formulated that the 'common' man considered physical pleasure, wealth, and social standing as the highest goals in life. He emphasised that all people are looking for happiness and that for most people this happiness was realised by physical pleasure, wealth and social standing. Aristotle called this the "Pleasure Principle".

That does not mean to say, however, that Aristotle necessarily agreed with such means of achieving happiness. On the contrary, he believed that real happiness could only be achieved by means of contemplating eternal, unchangeable values that he called 'essences'. His formulation of the principle of happiness was more of an empirical observation of the activities of the common man, whereby he concluded that such activities were being conducted with the aim of achieving happiness. These observations by Aristotle have been acknowledged over the centuries by philosophers until today. Indeed, the Pleasure Principle became the foundation of a philosophical school called utilitarianism. The founder of utilitarianism was the British philosopher Jeremy Bentham (1748-1832). Bentham developed an aesthetic system that was entirely based on the principle of happiness. According to Bentham, nature has placed man under two masters—pleasure and pain—and all human actions are focused on obtaining pleasure and avoiding pain.

John Stuart Mill

The 19th century British philosopher John Stuart Mill (1806-1873) further elaborated on the ideas of Bentham, but he emphasised that maximizing happiness is the ultimate goal of each and every person. Mill distinguished between various forms and degrees of happiness, and he divided them into higher forms of happiness and lower forms of happiness. He actually distinguished between pleasure, joy, and happiness, while Bentham regarded all forms of happiness as basically the same. Philosophers, sociologists and psychologists do not really differ in their opinions about the general human endeavour for happiness and good fortune. There are, however, some differences of opinion regarding the definition of happiness, and, as might be expected, many more differences when it comes to defining the way that happiness can be achieved.

Will power

Desire represents the active function of consciousness and is mainly driven by emotion. Will power, on the other hand, is the transformation of these desires into action whereby both emotion and intelligence play a part. If someone desires fame and success in politics, for example, then he will transform that desire into a number of specific actions, by means of will power. But if he wants to be successful then he has to use his intelligence and knowledge and make well-considered decisions and act in a calculated way. He has to take all sorts of circumstances into account, such as his political competitors, the media, politics, his financial means, etc. The attempt to achieve one's desire or ambition is often a long road in which will power and perseverance are constantly tested. Moreover, one's analysis and understanding of the subject and all other relevant factors has to be accurate. Will power without the necessary support of knowledge will almost inevitably lead to failure.

At the deepest level will power is a characteristic that cannot be reduced to something else, just like intelligence, emotion and desire. The famous German philosopher Arthur Schopenhauer (1788-1860) made this well-known statement on will power:

"Man can do what he wants, but he cannot want what he wants."

Sometimes this statement is also translated as 'man can do what he desires, but he cannot want what he desires.' Schopenhauer believed that will power and desire were two sorts of interchangeable attributes. The meaning of this statement is that we can control our actions, as far as they are an expression of our desire and will power, but we cannot control the desire

Arthur Schopenhauer

itself and we cannot control **what we** desire. Schopenhauer's observation was correct. Desire and will power are fundamental functions of consciousness that cannot be reduced to something else or to something deeper.

Will power has the ability to generate action without cause, the only cause is will power itself. Consciousness, by means of will power, can therefore literally create, and no other substance or reality possesses this characteristic. This fact is of major importance in understanding the origin of power, energy, movement and change in the universe.

Self-awareness

Lastly, consciousness has a unique characteristic, namely that it is aware of its own consciousness. This also seems, just like so many matters referring to consciousness, to be a matter-of-course. However, upon closer look, this is not the case at all. Self-awareness is one of the clearest expressions of the principle of oneness-in-multiplicity, for self-awareness represents the I, the self, the identity of the observer. All observations, sensations, memories, feelings, and expectations come together in one point, in one entity, which is the conscious self, the I. What the I is, is in the end impossible to explain, it just IS. It cannot be explained by referring to something else. The I is so unique and fundamental that it can only be understood by referring to the experience of I. Fortunately we are all conscious individuals and I's, so we all know very well what we are talking about. To cite Haisch again, knowledge of consciousness, and therefore also knowledge of the I, or the Self, is elementary and just as inalienable from my existence as water is inalienable from the ocean.

What is consciousness?

But what is this mysterious phenomenon, consciousness? We only know it because it is there, but one could say this about everything. What is matter, what is mass, what is gravity, what is the colour blue, what does sweet taste like, or what is consciousness? All these things are, in the end, just what they are. The difference between consciousness and all other things is that it is only because of the existence of consciousness that these other things become visible, amongst which is consciousness itself. From this point of view, consciousness becomes very important indeed. In fact, one prominent philosophical movement states that consciousness is one of the most important energies and features of the universe, just like mass, gravity, radiation, time, and space. This movement is called idealism, a movement that we have discussed before and that regards consciousness as the foundation of existence. Idealism as a philosophical movement assumes that ideas—or consciousness—form the substance that reality is made of. The term idealism is mainly used in Western philosophy but does not completely explain what it stands for. The reason for this is that idealism is often used in Western philosophy in order to contrast it with realism, the notion that the world and the objects in it are real and lead an independent existence.

The Vedanta school avoids this contradiction between idealism and realism and between subjective consciousness and the objective world. The Vedanta school states that both consciousness and the objective world of matter are real, but that matter is a transformation of the energy that we call consciousness. Consciousness is defined here as pure energy and forms the irreducible foundation and substratum of existence. According to the Vedanta school, energy and consciousness are identical. In this vision consciousness is not merely an attribute of reality, but it *is* reality. The Vedanta doctrine calls the origin and primal energy of the universe *Brahman. Brahman* is defined as *sat-chit-ananda*, which means 'eternal existence, made of pure consciousness, infinite knowledge and bliss'. This vision puts reality in a new perspective and elevates consciousness from a side-effect to a core reality. Energy can be influenced and changed, including it being converted into mass, but it can never be destroyed. Therefore since mass and energy cannot be destroyed, they are eternal. Since consciousness is identical to energy, it also cannot be destroyed; it is an eternal and inherent part of existence. Consciousness, therefore, is the foundation of reality and constitutes the bases of the energy that reality is made of. This fact is one of the main themes of this book and will be further substantiated in the following passages and chapters.

The mystery of the paradox

In previous chapters, we have frequently referred to the fact that reality is endowed with some most special and mysterious features. These 'special and mysterious features' are expressed in the fact that apparently everywhere, at every level of existence, we can observe irreconcilable contradictions—the so-called paradoxes. In chapter 5, 'Defects of materialism', we referred to some of these paradoxes when discussing particles, physics, and especially quantum physics. We discussed one of the core findings of quantum physics, which is that at a subatomic level, particles behave like fields of energy and vice versa. This is a fundamental paradox that has shaken the world of physics. Later in this same chapter, we also discussed the more philosophical paradoxes of Zeno. These paradoxes refer to the fact that time and space manifest themselves both as a unity or continuum, as well as an infinite multiplicity of infinitesimal yet separated points. In chapter 9, in our discussion on the different complexities of existence, we also discussed simultaneous movement and permanence, a paradoxical phenomenon that manifests itself everywhere in the universe. It particularly manifests itself within the willpower faculty of consciousness.

For the remainder of this chapter 10, we will discuss the paradoxes that are integral features of consciousness. Subsequently, in chapter 12, we will describe how paradoxes constitute the foundation of reality and how they form the core of the very being of God, and how these paradoxes particularly manifests themselves through the functioning of our own consciousness. Indeed, in describing the attributes and functions of consciousness earlier in this chapter, we have become aware of the truly mysterious, mystical, and incomprehensible nature of consciousness. Yet there remains one characteristic of consciousness that beats all other attributes in terms of its level of mystical incomprehensibility, and that is the ability of consciousness to unite opposite and paradoxical attributes. This paradox of opposite attributes can be divided into the following four areas:

- Simultaneous oneness and manyness,
- Simultaneous change and permanence,
- Simultaneous fullness and non-fullness,
- Simultaneous action and inaction

Oneness and manyness

The oneness of consciousness is represented by the I, the Self, the centre of consciousness. The manyness of consciousness is determined by the multiplicity of observations, sensations, thoughts, memories that occur both sequentially and simultaneously. The essence, the core of conscious observation, is that this multiplicity converges in and is concentrated in a unity, the Self, simultaneously and in the same place. Consciousness overarches and coordinates this multiplicity, this manyness and brings them together in a conscious experience. According to Vedanta doctrine, consciousness is not a static phenomenon, neither quantitatively, nor qualitatively. Consciousness is able to increase in intensity and expand quantitatively both in time and space. An increase in knowledge and insight entails the simultaneous expansion of consciousness in space and time. This point has been elaborated on extensively in the previous section of this chapter. The simultaneous unity and multiplicity of space and time and the simultaneous unity and multiplicity of consciousness are the foundation of what we call comprehension and understanding. They are crucial ingredients of what we call consciousness and what we call conscious experience. These attributes make consciousness absolutely unique and incomparable. It also largely determines the distinction between consciousness and conscious experience when compared to the Artificial Intelligence of a computer. Apart from this difference, a computer does not have conscious experience of the computations that have to be performed, and neither does it have emotions or a sense of identity.

Change and Permanence

Apart from the simultaneous oneness and multitudinous characteristics of consciousness, consciousness also manifests the paradox of simultaneous change and permanence. Regardless of the amount of different observations that fill our consciousness and the continuous change of that observation, our consciousness itself remains unchanged. While I sit here typing hundreds of thoughts go through my mind and I am simultaneously touched by thousands of small and large impressions. Nevertheless, I simply remain who I am. My feeling of self, of unchanged indivisible identity, is always present. Therefore, I am able to stay a unity under this simultaneous bombardment of various impressions and sensations. I can also maintain this unity in a sequence of countless impressions, sensations, and thoughts in time. I can understand and process these thoughts and impressions, since I can relate the previous thought to my current thought that I can again link to a future thought. Without this underlying identity, continuity and permanence—both in space and time—understanding and insight would be impossible.

Fullness and emptiness

Furthermore, the paradox of simultaneous fullness and non-fullness is also a crucial characteristic of consciousness. Our consciousness is constantly filled with sensations, similar to an empty container that can be filled with objects. The difference is that the container of consciousness is not empty; it is actually filled with consciousness. Despite the fact that the container is filled with consciousness, it can still allow itself to be completely filled with sensations and experiences. This is again a paradoxical, incomprehensible combination of attributes even though it refers to that which enables us to understand things, namely consciousness itself.

Action and inaction

Finally, the paradox of simultaneous change and permanence also refers to the active attributes of consciousness, in particular will power. Will power literally implies action without cause, action initiated by the I, the Self. While action implies change, the I remains unchanged; it is action within inaction. While the I generates the energy that is necessary to execute the action, the I does not lose energy. This characteristic too is of a paradoxical nature.

In conclusion, the paradoxical attributes of consciousness give it unique and powerful abilities. The unity and multitudinous function of consciousness, together with the change and permanence function, make it possible for consciousness to initiate, create and organise. Through conscious interaction, separated and independent objects can be moulded and organized into one organic whole. This organic nature of consciousness is of major importance and we can experience this ability through our own thoughts and consciousness. In the previous paragraphs, we used the example of writing a book to indicate how our consciousness processes and generates information. Consciousness is the only characteristic known to us that has such an organic ability in which unity and manyness come together in one single working structure. This ability is the foundation of what we call information and determines the definition of information. Information is the structured coming together of manyness in an organic unity. The binding and coordinating force within such an organic unity is consciousness, from which design and organic structure are created. The universe, as we have concluded before, is filled with information and filled with organized, structured complexity. It is consciousness that is the source of all creative and creating abilities and also the source of what we call information.

The core and definition of information is the consciousness of a multiplicity of things within a unity and a unity within a multiplicity. The number 2 only has meaning if the number 1 exists, if the number 3 exists and if all the other numbers in the set of natural numbers exist. The unity aspect of numbers is that they belong to the category or set of 'numbers', that they share characteristics and that therefore they are comparable. The well-known proverb 'one cannot compare apples to oranges' makes sense in this respect. The multiplicity of numbers is due to the fact that it consists of different numbers —1,2,3,4…..etc., where each number is different and unique. We accept this contradiction as a matter-of-course; as a self-evident and obvious fact, and we often do not think about it any further. Yet multiplicity in unity and unity in multiplicity it is deeply mysterious and paradoxical reality that forms the core and definition of what we call 'information'. It also forms the core of our consciousness; it enables a multiplicity of impressions and sensations to converge at one single point simultaneously. That point is me, the self that is conscious of the multiplicity of things, that does not lose sight of the connections between the multiplicity of things, and that encapsulates everything from a coordinated and concentrated centre.

Personality

All of these attributes and functions of consciousness converge in what we call personality. Personality is, in essence, determined by the presence consciousness, but it is the specific, individual manifestation of consciousness with all its different attributes and functions. Consciousness is also the foundation of what we call 'life'. Everything stated above about consciousness is integral and applicable to the concept of 'personality'. Personality is consciousness and consciousness is personality; these are synonymous concepts. Consciousness without personality is unimaginable in the same way that the sun is unimaginable without light and heat. Personality is the way in which consciousness manifests itself to others and to other conscious beings. Personality is also the way in which consciousness manifests itself to itself. We are personalities and that is also how we perceive ourselves. Personality is a non-reducible reality in the same way that consciousness is a non-reducible reality. And personality is, just like consciousness, infinitely complex, but at the same time simple. In the following chapters it will be further discussed how energy, consciousness, and personality are the foundation and source of reality, and how energy forms the substratum of reality, the substance of which everything is made. While energy possesses infinite actual and potential attributes, the most dominant and all-determining characteristic is consciousness, and consciousness manifests itself through that which we call personality. Energy, consciousness, and personality therefore form the ultimate and non-reducible foundation of reality.

"I do not feel forced to believe that the same God that equipped us with senses, reason and intelligence, intended for us not to use them."

Galileo Galilei

The scientific foundation for the existence of God

At this stage it is useful to provide a brief summary of the principal conclusions and arguments that have been presented so far. The most important conclusion based on the teleological and anthropic principles is that God must indeed exist: the complexity of the universe especially in relation to the creation of the four fundamental forces of the universe, their values and interactions, at the moment of the big bang, cannot have occurred by chance or chaos. It is statistically impossible, and the only alternative explanation is that the universe must be led by intelligence. In subsequent chapters, we also concluded that matter is not the only reality and that there is an energetic reality apart from matter. Indeed, we concluded that energy is actually the foundation of reality in which matter is a derivative transformation of energy, and that both the material as well as the energetic realities possess extraordinary attributes that are

In previous page: the world-famous painting by Michelangelo in the Sixtene Chapel of the Vatican: an image of God while he is creating the creation. In this image God creates the first man, Adam. Michelangelo depicts God as a handsome, vital old man with a majestic, grey beard. Many of the Christian images of God go back to Genesis (1:27): "So God created mankind in his own image, in the image of God he created them, male and female He created them." Monotheistic religions in general regard God as a person, albeit as a person with transcendental characteristics. In the Vaishnava Vedanta doctrine the transcendental personality of God is depicted very explicitly, both in images as well as in descriptions. Like the Christian version, God is also described as exceptionally handsome, however, God is also described as young, charming and full of vitality. In the Vedanta school, God is described as Bhagavan, possessor of the six most attractive characteristics in infinite abundance: beauty, intelligence, detachment, fame, riches, and power.

supernatural in a natural way. Furthermore, we concluded that the universe is a unity, or a multitude within a unity and a unity within a multitude.

These conclusions are mainly based on hard scientific and mathematical arguments. There has also been a more philosophical line of reasoning, wondering why there is something rather than nothing and that if there is something, then why it is so complex? The logical answer to these last questions is that if there is something then it must be meaningful and have a purpose. If it did not have a purpose, as atheism suggests, then existence itself would be totally illogical and inefficient. In a universe that is governed by the laws of nature and where the principle of effectiveness plays an important, if not crucial, role it is very illogical that this effectiveness should be created by a principle that represents the complete opposite. All of these arguments and discussions led us to the reasonable and logical conclusion that God does indeed exist; a conclusion that can be quite confronting to many. From our limited human perspective it seems very unreal and improbable that there exists, after all, an infinitely intelligent and conscious being with whom we are personally connected and who is completely aware of everything in infinity and eternity. It seems improbable, for reality appears to be so different: so imperfect, arbitrarily chaotic, mundane, fleeting and full of random suffering. The existence of God appears to contradict our day to day experience of reality and moreover, he appears to be completely hidden from our everyday lives. In the words of Richard Dawkins previously quoted in Chapter one:

> *"The universe we observe has precisely the properties we should expect if there is, at bottom, no design, no purpose, no evil no good, nothing but blind, pitiless indifference."* [87]

The ability of God to remain hidden is what is called in Vedic texts the power of Maya, the cosmic illusion that enables a living being to believe that he is God or a little bit like God. In order to delude oneself that he is God, the 'real' God has to make himself invisible. There is no doubt that *Maya* has succeeded brilliantly in that. These issues relate to the problem of the Theodicy, which has been discussed at length in previous chapters.

God is supra-rational and incomprehensible

There is, however, another point that deserves attention. From a strictly scientific and empirical point of view, neither the existence nor the non-existence of God can be proven despite the fact that throughout history there have been people who claim to have had a direct experience of God. The mystical experience that we referred to in chapter 7 can certainly be considered as a direct and empirical observation, however, it requires other instruments and other forms of consciousness to be obtained. Therefore, a mystical experience is not directly accessible to all as empirical evidence. On the other hand, the scientific proof that is accessible to all as has been explained in this book, is a negative proof. This means that it has been demonstrated, especially founded on cosmological and mathematical arguments, that the non-existence of God is impossible. The organized complexity of the universe is so massive that the only possible explanation for that level of organized complexity is that it was generated by intelligence and information, whereby coincidence and chance are reduced to mathematical impossibilities. Therefore, if there are only two options to the solution of a problem, and one of those options turns out to be impossible, then the other option has to be true, based on the process of elimination. A positive proof of God's existence would for example constitute his appearance in the firmament where he became visible to all on earth in this way revealing himself directly. If he would subsequently address all of mankind in perfect English and three or four other world languages with a pleasantly thundering voice, stating who he is, the proof would be direct and undeniable. If on top of it he would perform some additional miracles, such as hiding the sun behind his colossal appearance without disturbing the climate on earth, his existence would be an undeniable and absolute fact. If he then gave us the time to record and document his appearance with our smartphones, cameras, radars and satellites, to keep everyone convinced that this was not a collective hallucination the definitive scientific proof would have been delivered. From then on the debate on his existence or non-existence would have been settled forever.

Hypothetical questions aside, in our current situation there is no such direct positive proof, either for God or for the non-existence of God. Indeed, the universe is so mysterious and incomprehensible that both the atheistic and theistic explanation are incomprehensible to a certain extent, and they challenge the boundaries of human understanding. When scientists talk about singularity and the big bang, we are actually talking about an unimaginable reality. The documentaries on Discovery Channel about the origin of the universe, starting with singularity, via the big bang up to the formation of the first star systems and solar systems are always accompanied by smart audio-visual effects and computer simulations. The audio effects accompanying the images of the big bang always depict an actual enormous explosion, followed by many more explosions that are even more impressive. The intention of all the effects is to make an event that is totally incomprehensible for humans somehow conceivable. The same holds for God; each description of God defies human imagination. Whilst we hardly ever come across documentaries about God creating the universe on the Discovery Channel, if we were to see such a documentary, then it would be filled with impressive effects in order to make an incomprehensible reality somehow conceivable. The point is that both God and singularity are fundamentally incomprehensible. Considering the incomprehensible nature of both worldviews, we have to choose the worldview that matches our experience, and especially scientific knowledge, the best. At this point the conclusion is absolutely clear; to cite another famous physicist, Dr. Bruce Gordon:

> *"When the logical and metaphysical necessity of an efficient cause, the demonstrable absence of a material one, and the proof that there was an absolute beginning to any universe or multiverse are all conjoined with the fact that our universe exists and its conditions are fine-tuned immeasurably beyond the capacity of any mindless process, the scientific evidence points inexorably toward transcendent intelligent agency as the most plausible, if not the only reasonable explanation."* [88]

God as scientific reality

As has been demonstrated so far in this book, there are powerful and convincing scientific arguments that support the existence of God. That being the case, we really will have to get used to the idea. However, this might be somewhat difficult for most of us because God, at least as he has been presented by most religions throughout the centuries, is actually akin to some kind of fairy tale. A fairy tale that we, when we are religious, accept as a reality, but that nevertheless always has a fairy tale-like feel to it. God anchored in rational thought based on scientific insights and evidence is quite something else. Science represents the actual and real world, the world of facts and matter; the world we experience. We have currently accepted that even scientific facts are often beyond our direct experiences, as we have concluded previously. But, nevertheless, they remain hard and actual things, because we know that under certain conditions they can be made visible, and as such we know and accept that they exist. The atom is an example of this; we cannot directly perceive an atom, but we know it is there, for we experience material objects which we know are made out of atoms. We also know the atom exists because of the type of power it can generate. Fortunately, few people have experienced this power personally. On the other hand, almost everyone has seen a movie or documentary about the explosion of a hydrogen bomb or atomic bomb. When we watch such a documentary we know that it is real and that on 6 August 1945, an actual atomic bomb exploded above the city of Hiroshima. We have witnessed the horrific consequences of the atomic bomb on film and the unmeasurable human suffering that it can cause. Therefore, we know that it is real, that it actually happened and that it is part of our actual reality.

In order to place God in such a scientific frame as being a real and actual part of our everyday reality is puzzling and shocking. The existence of such a being is clearly beyond our imagination, and contradicts in many ways our everyday experience. However, many things which we know to exist, exceed our imagination. Not only that we cannot see the atom, but we cannot even truly imagine its size, structure and speed. Nor can we imagine the size and structure of our Milky Way galaxy with its diameter of 100,000 light years. In this book, we have frequently highlighted the unimaginable and supernatural nature of reality uncovered by scientists over the last few hundred years. Despite that, we accept that these unimaginable things really do exist. It is a useful exercise to put the existence of God into a similar perspective. For instance as an interesting thought experiment we can take a look at our own consciousness and position in the cosmic hierarchy of living creatures and things. From the perspective of an insect, for example, what we do as humans is unimaginable. Everything I can do as a person is totally incomprehensible for an insect and above and beyond any type of limited understanding that it possesses. The fact that I can crush the insect with the tip of my finger, while typing on my laptop or having a telephone conversation, is beyond the imagination of an insect. From the perspective of the insect, I might as well be an all-mighty Deity, with unimaginable power and intelligence. Our own consciousness is, of course, extremely mysterious and incomprehensible, even for us, let alone for an insect. When we extend this perspective in the opposite direction, then we can also imagine that there is a being that possesses a level of consciousness, intelligence and power of a magnitude that is far greater than ours, and which far exceeds our imagination. We in turn become the insect like creature, —which is not flattering,

I know— in comparison to such a superior being possessed of such superior and unimaginable abilities. Such a perspective is entirely feasible and logical. We sometimes see such creatures in science fiction or horror movies; knowing that it is only a movie. However, science currently shows us that such a being, God, actually does exist and that this is not a movie.

God has a burdened history

In the course of human history God has represented the highest of the high and the holiest of the holy in religious traditions and has been treated with the utmost respect. Yet God has also remained largely incomprehensible due to our limited knowledge and consciousness. Unfortunately, it is exactly this incomprehensibility linked to a deep-rooted respect, awe, and even fear of God, which has often been the foundation for all sorts of ignorance, superstition, wrong assumptions and even evil. Awe and fear reduce the margin of error to zero; in the face of God no errors are tolerated; neither errors in the interpretation of the spiritual ideology nor errors in the application of the ideology. After all, the stakes—eternal damnation for every wrong choice—are too high. This compulsive attitude leads to intolerance and aggression and often eliminates the positive elements of the ideology, such as compassion, tolerance, humbleness, and servitude. Furthermore, this compulsive attitude becomes even greater when 'religion' is accompanied by anti-rationalism and a hostile attitude towards science and reason. The combination of superstition, awe, and anti-rationalism is an especially toxic mixture that has caused tremendous misery throughout history. These phenomena appear in all sorts of shapes and forms and in all religions and spiritual ideologies. The history of religion has therefore demonstrated a controversial and burdened picture of individuals and institutions armed with the word of God fighting for power and influence. Because of ignorance, dogma, sectarianism and superstition, many bloody religious wars have been fought in the name of God. Therefore, it is important, in fact, it is essential, to analyse and understand the concept of 'God' within the definition and real meaning of the word and not just as a historical and emotionally burdened concept.

It is beyond the scope of this book to give a complete overview of the various world religions and the manner in which God is described in these religions. It is clear, however, that God is the central theme in every religion, and that every religion aims to understand God and aims to understand how our earthly existence and how this material universe is connected to God. This chapter will focus on the relation between religion and human reason in general and the relation with philosophy and science in particular. These two—religion and reason—have had a difficult relationship within many of the religious traditions. By rejecting human reason in favour of strict dogmatism, many sectarian conflicts have arisen due to theological differences of opinion. Because of such strict, emotionally charged dogmatic attitudes, these differences could never be expressed, debated or substantiated on the bases of logic and evidence. Rather they were suppressed, in many instances with the use of brutal violence.

Religion and irrationality

This has especially been the case with the Western religions of Christianity and Islam and has led to numerous conflicts, in some cases violent conflicts that were totally unnecessary. Such conflicts caused tremendous suffering, and definitely did not bring us closer to the essential goal of all religions—namely the elevation and greater happiness of humanity and society and the enhancement of human knowledge, not only in this world but also in the next. It has also vastly stained religion in general and the concept of 'God' in particular. The cause of this can mainly be found in the roots and source of a religion, or in the shape of the early phase of its development. Anti-rationalism in Christianity, for example, is an inheritance of the early Christian culture with its limited philosophical and intellectual foundation. Christianity was founded—from the beginning—on the revelations from the

Bible and the life and lessons of Christ, with a strong emotional overtone. The gap between religion and reason is a common theme in the history of Christianity and it manifested itself in many different ways. The Christian theologian and Church Father Tertullian (155 – 200 AD) is an example of a person who formulated a theory about the relationship between faith and reason. In the 2nd century he said that proper knowledge could only come from religion, and that religion and reason excluded each other. His proposition was: "Credo quia absurdum est", or in other words, "I believe because it is absurd." Martin Luther (1483 – 1546), the influential German theologian and founder of the Protestant Reformation, said that reason was 'the whore of the devil.'[83] The famous 19th century Christian theologian and philosopher Soren Kierkegaard (1813 – 1855) described his philosophical ideas as a 'leap of faith', or a 'leap of the absurd'. Clearly, it appears from such statements that, religion and reason did not go well together.

However, it is certainly not the case that all Christian theologians shared this opinion. It is a known fact that the Greek philosophers had a major influence on the development of Christianity and Christian theology. Plato, Socrates, Aristotle, and Plotinus had a significant influence on Christian thought and the development of Christian theology. Many Christian thinkers were not negative about the philosophy and logic provided by the Greek philosophers and which they tried to incorporate into the Biblical vision of the world and reality. The previously cited Thomas of Aquino is a good example of this. Another example is the famous Church Father Augustine, who lived from 354 to 430 AD. He was a theologian and philosopher who was strongly influenced by the Greek philosophers Plato and Plotinus and is also sometimes regarded as the most influential Church Father with respect to the formation of Orthodox Christian theology. However, Greek philosophy was nevertheless alien within the Christian context, having been imported from another culture; nor was it derived from the Holy Scripture, and eventually this led to conflicts. The former cited Tertullian said: 'Free Jerusalem from Athens and the church from the Academy of Plato'. Despite the philosophical, rational influence of Greek culture, anti-rational movements in the history of Christianity have had a major influence.

Galileo

The battle between reason and religion has led to numerous historical conflicts between the church on one hand and scientists and philosophers on the other hand. The persecution of Galileo Galilei (1564 – 1642 AD) is without doubt the most well-known of these. Galileo contradicted the theologians and biblical scholars by claiming that the sun is the centre of our solar system instead of the earth. His theories and discoveries were founded on those of Copernicus (1473 – 1543), the famous Polish astronomer. He concluded, based on his observations and calculations, that a heliocentric theory of the universe was a much better and simpler explanation for the movement of the heavenly bodies than the Ptolemaic or geocentric explanation. The geocentric vison of our solar system was propounded by Ptolemy in the second century. He had based his theories in turn on Aristotle, and these were later adopted by the church.

Geocentrism was perfectly aligned with the Biblical explanation of the origin and structure of the universe as explained in the book of Genesis. According to the Biblical narrative, earth was the central point of the creation of God, and therefore the centre of the universe. Earth was believed to stand still, while the heavenly bodies—amongst which is the sun—were believed to rotate around earth. Galileo was the first to study the heavens with a telescope and he gained evidence for the statement of Copernicus that earth, in fact, rotates around the sun and was not the centre of the universe. At that time, such a change from geocentrism towards heliocentrism was a major revolution in the understanding of the way the heavenly bodies worked and is known amongst historians as the Copernican revolution. This revolution threatened the authority of the church however, and the church decided to fight back. As a result, Copernicus never

Galileo talking to some church scholars, where he explains his findings about the structure of the solar system and the heavenly bodies. Galileo's point of view was very controversial and, in the end, led to his arrest as part of the inquisition.

officially published his ideas; he only discussed them in private. Indeed, his most important work was only published on the day he died. Copernicus therefore spread his ideas quietly, but Galileo was far more outspoken. His powerful and well-thought-through writings and propositions made him a serious threat

Galileo Galilei

in the minds of certain people in charge of the church. He was ultimately convicted by the inquisition and forced to recant his opinions and spend the last eight years of his life in his own home under house arrest.

Bruno Giordano (1548 – 1600) is also a well-known historical example. His fate was even worse than Galileo's for he was burned alive at the stake. His terrible crime was to state that the universe was not geocentric, and that space was not finite, but infinite. The church regarded his theories as heresy and the inquisition sentenced him to death. What mattered to the church and Christian scholars was that the power of revelation and religion was more important than and superior to reason and empirical observation. People like Tertullian even regarded the power of reason as the enemy of religion and as a negative force. This theme has come back time and again throughout Christian history. It has made an indelible impression on how culture, philosophy, and science developed in Western society. The emphasis on religion and knowledge based on revelations over

The trial of Galileo in 1633. Galileo was convicted by the inquisition and ordered to recant his points of view regarding the structure of the solar system. He was placed under house arrest, a situation that lasted until his death in 1642. According to popular stories, Galileo was said to have mumbled after he had denied his proposition that the earth rotates the sun, "… and yet it rotates." From a historical perspective, this has been one of the biggest mistakes of the Catholic Church, for Galileo was demonstrably right. It was not until the second half of the 20th century, on the 31st of October 1992, that the Church, led at that time by Pope John Paul II, apologised for this fallacy.

and above reason and observation, created an anti-rationalism and intolerance in general which have had a negative influence on Christianity. Other religions as well, such as Islam, show strong anti-rational and sectarian tendencies, with little tolerance for people with a different view. Many wars have been fought with religion and ideology at stake. Rivalry between Christianity and Islam has a long history, from the crusades up to modern times. The Protestant division in Christianity that originated with Martin Luther in 1517 led to many bloody, sectarian conflicts within Christianity itself. Looking back at these events, with the knowledge we have now, we can only be amazed at the levels of stupidity. The consequence, however, is that religion—and in particular institutionalised religion—has lost a massive amount of credibility. Consequently, the credibility of God has also suffered.

However, it is important to detach God from this toxic history created by men who were often misguided or lacking knowledge at best and at worst driven by selfish pursuit of fame, wealth and/or power. Instead, we should be aware that a scientific and philosophically sound vision of God can exist. It should be a vision that matches the level of knowledge of modern man and the information society that we have created. It is important to distinguish the emotionally charged ideas about God from objective and scientific insights. Within this meaning even the term 'God' is of secondary importance. In the end, it is about the actual nature and attributes of the entity that occupies the position of the Supreme Being and who is the cause and maintainer of the universe.

The foundation of all religions

At their core, all religions of the world are correct. The most important and central theme of all religions is the existence of an infinitely intelligent supreme being, the existence of God. This is the universal truth that forms the foundation of all religions and at this point all religions agree with one another. However, within this commonly held truth there is nonetheless a range of different interpretations about who or what God is; from monotheism, pantheism, deism, monism, and up to polytheism. Moreover, the way in which God reveals himself, or how we should approach or honour Him, or how man should use his life knowing that he has to answer to God in the end, and the vision of how this should be done, differs from religion to religion. This also holds for the manner in which religion claims the exclusive truth and whether or not a religion is open to other sources of knowledge. Unfortunately, in all these areas history shows that religion and their adherents have failed. Intolerance and sectarian fanaticism have occurred more often than not, and they have caused a lot of unnecessary suffering. This does not need to be and should not be the case.

In the end, the process of discovering the truth requires that a convergence takes place between religion, science and philosophy. Whilst there will always be differences in interpretation and differences in nuance, these differences are mainly caused by the paradoxical complexity of reality. The *Acintya* principle of the simultaneous unity and multitude of God and His energies is an important explanation of this, as well as the fact that God is infinite and boundless. In this book, we have often referred to the principle and paradox of simultaneous oneness and manyness. Whilst this was discussed quite extensively in the chapter on consciousness, this discussion will be expanded upon in the following chapters. It will be explained how this principle constitutes the foundation of the energetic reality and, in particular, comparisons are made with the fundamental principles of modern physics. Thereby, we reinforce the foundations for a scientific approach to God that is compatible with reason and empirical observation, as well as a philosophical foundation that establishes the unity of God filled with infinite variety.

Chapter 12
THE PARADOX; THE FOUNDATION OF REALITY

"My own suspicion is that the universe is not only queerer then we suppose, but also queerer than we can suppose."

J.B.S Haldane in *Possible Worlds*

"Something unknown is doing we don't know what"

Sir Arthur Eddington,
in a statement about quantum physics in 1928

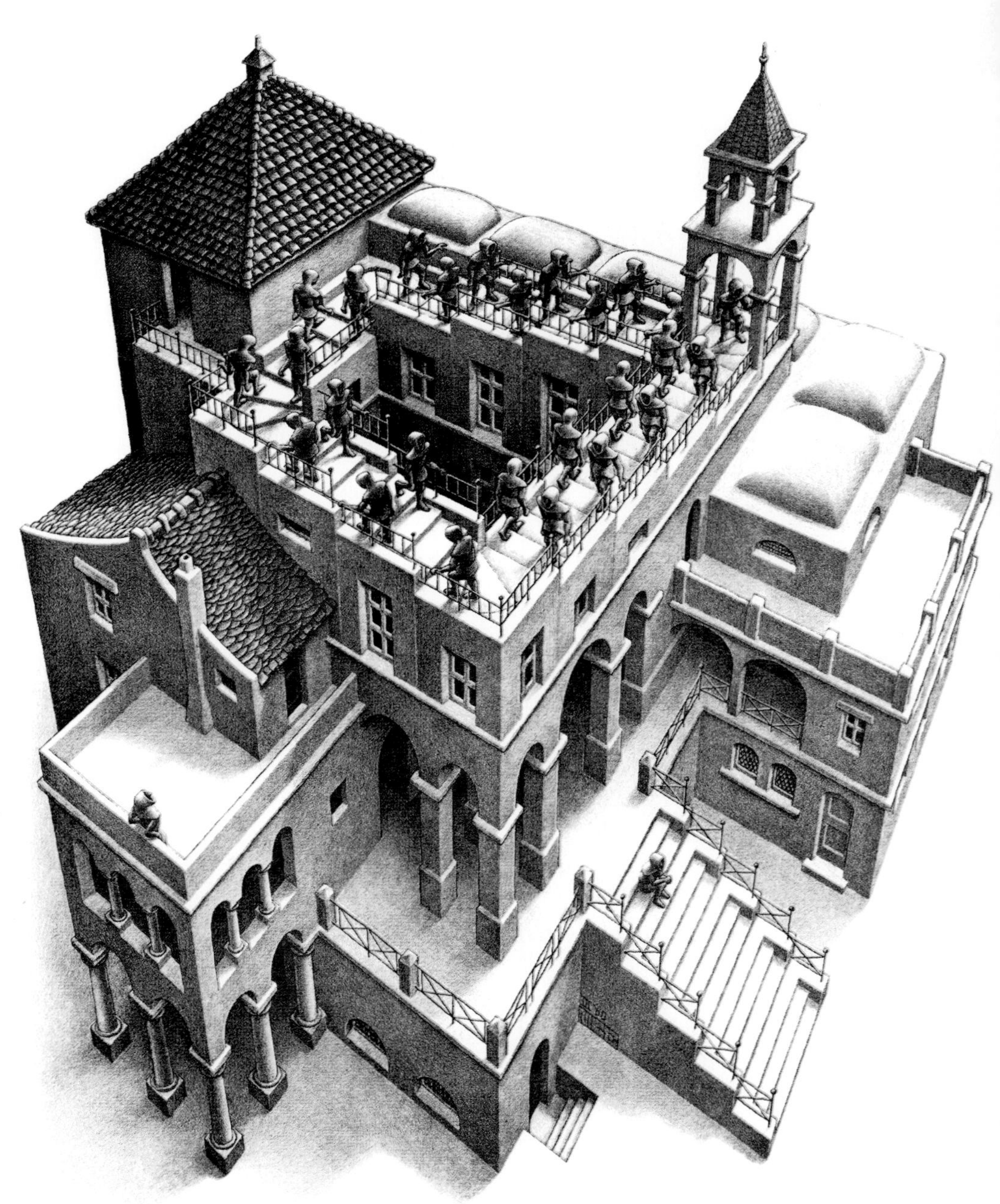

The energetic reality

The existence of God has major implications with respect to our understanding of the universe, both scientifically and philosophically. Our understanding of matter and energy and the functioning of these realities within the cosmic order take on a whole new meaning. The most important fact is, of course, the notion that the universe is governed by intelligence and that the universe has a purpose. In this chapter we will summarize the most important scientific consequences of the presence of such an all pervasive intelligence, which consequences were already discussed in more detail in previous chapters. The focus of the chapter, however, will be to offer further insight into the elementary mystery of the paradox that constitutes the foundation of reality and that represents the intrinsic nature of God.

God is the foundation and the source of energy

At the deepest ontological level God consists of himself, but for reason of clarity and practical use of language, we will define the substance that he is 'made of' as pure energy. He is pure energy, without boundaries or limitations, whereby energy manifests itself both as an infinite source, which overflows and expands

Previous page: the famous drawing of the world-renowned Dutch artist M.C. Escher. Escher was a master of creating conflicting perspectives, to be compared to the paradoxes we encounter in science and philosophy. The basic assumption is that when you look at the drawing consistently from one perspective, it is inevitable that at a certain point you encounter an entirely different, conflicting perspective. Regarding paradoxes, the laws of logic work in a similar way.

into an infinite manifestation of energy. This relation between energy source and energy is of crucial importance and is also at the core of the mystery of existence. In order to illustrate this, the relation between sun and sunlight provides a good comparison. The sun is localised and has a specific identity, but it is simultaneously the source of non-localised, all-pervading energy in the form of sunlight and radiation. Since sunlight emanates from the sun, it is effectively a part of the sun and shares similar properties with the sun. The sun includes the sunlight, both quantitatively and qualitatively, but quantitatively the sunlight does not include the sun. When a sunbeam enters my study through the window, it is obvious that it is not the sun itself that is entering my study. The relation between the sun and its energy is therefore a relationship of simultaneous oneness and difference, qualitatively one and quantitatively different. That also implies that there is a hierarchical relationship between the sun and the sunlight, a relationship of dependency. In a comparable way God, the creator, and his energy—the creation—are simultaneously one and different, whereby the creation is dependent on its creator. The relationship of simultaneous oneness and difference expresses itself in two distinct manners, 1) qualitative oneness and quantitative difference such as the sun and sunlight analogy, and 2) the simultaneous existence of indivisible unity and diversity, whereby God represents indivisible unity and his creation represents diversity. The principle of simultaneous oneness and difference, of simultaneous unity and multitude lies at the core of this relation and is summarised in the principle of *Acintya bheda abheda tattva,* as was discussed in the previous chapter. This theme will be discussed in detail in the following paragraphs.

One of the most important implications of the existence of God is that the universe is a fundamental unity. In previous chapters, we already concluded that nothingness and emptiness do not exist, which conclusion implies that the universe must be a unity. This unity consists of an infinite field of energy that in turn contains infinite chains and levels of connected and overlapping energy fields. This conclusion has far-reaching consequences. First of all, it explains why objects can influence each other over a certain distance. After all, the space between the objects is not empty, but filled with energy. Energy arranges the mutual influence and makes action at a distance possible. Secondly, this implies and confirms that chaos cannot exist, for chaos can only exist in the absence of unity. Unity is the opposite of pluralism, and pluralism is the foundation for chaos and randomness. However, unity does not exclude diversity, and can therefore best be described as an organic unity. Organic unity implies a cohesion and functional connectedness between all diverse things that are part of this unity. Diversity, as part of an organic unity, is therefore not the same as pluralism. Diversity is inseparable from unity, just as individual, separate numbers are only meaningful because they are part of the unity, which is the set of numbers. The number 5 is only meaningful because it is preceded by the numbers 1, 2, 3, 4 and followed by the numbers 6, 7, 8 and so on.

Finally, this organic unity is conscious and intelligent, so that the connected diverse things are indeed functional through the infallible and inconceivably complex structures such as we witness in this universe. This the essence of the teleological principle, and forms the indubitable evidence for the existence of God as explained in chapter 2.

Energy is the base and the foundation of reality

We have also previously concluded that recent insights and discoveries in the field of physics assign a crucial role to energy. In the theistic universe, energy plays an even more important part; especially given the inadequacies of a purely mechanistic, materialistic explanation of the universe. It appears from many of the scientific discoveries and philosophical deliberations that energy has exceptional characteristics. The most outstanding feature of these characteristics is the lack of any limitation, and the full presence of all that is possible and even (nearly) impossible. Energy is the unbounded field of infinite possibilities, the reservoir of all actual and potential properties. Energy is that what generates motion and makes objects move, but it simultaneously constitutes the mass that objects are made of. While energy is full it is simultaneously infinitely fluid and flexible. This makes energy and the energetic reality an incomprehensible and supra-rational reality that transcends the materialistic, mechanistic view of the universe. However, mass and matter themselves also possess extraordinary characteristics, for they are basically a transformation of energy. These extra-ordinary qualities have been unveiled by the great scientific discoveries of the 20th and 21st century and have been discussed extensively in previous chapters. While energy and matter share their extra-ordinary nature, in the world of mass and matter, reality is organized differently. Due to the atomic particle structure, the different properties that co-exist within the energetic reality, become disbursed into separate particles, each possessing different properties. To use these properties for example by combining them, spatial constructs are required. Hydrogen and oxygen atoms have different properties when they are spatially separated, however when they become spatially connected their individual properties are combined and merged, and a new particle, a water molecule is created with entirely new properties. Thus space enters the equation to facilitate the required spatial construct. And time is also required since movement is needed in order to move the particles from separation to being combined. That in turn requires energy and work, and hence in the world of matter time and space become limiting factors. The world of matter is in a sense a digital representation of the world of energy, whereby the atomic packages simulate the underlying energetic reality. The material world is effectively a virtual world that exists within the real world of pure energy. This is exactly the definition of illusion, or *Maya* as is explained in the Vedic texts.

In our modern world, driven by electronics and computers, digitizing information has been a major advancement in technology and efficiency. By digitizing information, where information is converted and stored in small data packages, we are able, with the assistance of ultra-fast microprocessors and

340

Scene from The Matrix. We see an illusionary world, created by a computer code.

massive bandwidth, to manipulate this data at an incredible speed. In almost every way, the digital transmission and processing of information will outperform the slower, analogue way of communication. The digital world is a binary representation of the analogue, real world that makes it possible to simulate this real world. In a comparable way, the material world, constructed with the use of atomic packages, is a sort of digital representation of the world of pure energy, the spiritual world; with the difference that digital does not imply 'improvement', but rather a limitation. The 'atomic' digital world was created to create an illusionary reality that is permeated by all kinds of limitations. These limitations are created by the individual and collective karma of the living beings, the inhabitants of this atomic digital world. Within this context, it is interesting to see how filmmakers have used a comparable theme to portray this world as a massive illusion. The movie The Matrix is a good example of this. In this movie our world is portrayed as a computer illusion created by evil geniuses. Some heroic individuals try to escape this illusion, resulting in dramatic, Hollywood-style fight scenarios and spectacular computer simulations.

Many of these fights are therefore not normal fights but fights against computer generated illusions with the movie heroes struggling not to get caught up in an endless string of illusions. These and other similar movie themes did not appear completely out of the blue. Quantum physics in particular, and certain interpretations of quantum physics that we have referred to before, suggest that this universe is an illusion created by individual observers.

So far throughout this book we have demonstrated and discussed in detail the fact that reality, consisting of energy and matter, possesses extraordinary characteristics. In fact, energy manifests itself in a way that contradicts the world of mass and matter. The discoveries of quantum physics, relativity, and the anthropic principle lift the veil ever so slightly of the extraordinary and mysterious world of energy hidden behind the world of mass and matter. These discoveries confront us with a universe that challenges human intellect and present a reality of unimaginable characteristics and proportions, where concepts such as natural and supernatural are blurred and simply fade away.

Energy contains the paradox of simultaneous unity and multitude

We previously substantiated that the origin of the universe does not, ultimately, need to be simple. This was in response to the atheistic proposition that the origin of everything must be simple. At a deeper level, however, both propositions are correct. God is indeed ultimately simple, for he is one, but at the same time he is many, and contains a diversity of infinite complexity. God is therefore simultaneously one and different, one and many, a unity and a multitude, both of which exist next to each other and within each other. This is the most fundamental paradox which is the foundation and origin of all other paradoxes. Energy seamlessly unites these paradoxes; paradoxes that cannot be united in the world of matter and mass.

This principle of simultaneous oneness and difference, of simultaneous unity and multitude is the supra-rational foundation of reality, the basis of creation that manifests itself at all levels of reality. This is the core of the *Acintya* principle, and it is the core of the mystery of reality. In the chapter regarding consciousness, we already conducted a thorough analysis of this. At the deepest level consciousness and energy are the same, and therefore their attributes are exchangeable, and they also overlap. This analysis concerning paradoxical characteristics is therefore fully applicable to the description of energy, as will be discussed in this chapter.

But what are paradoxes? What are their characteristics, and what are their effects on our understanding or reality?

The paradox

Many paradoxes have been discovered and recorded in the course of history. Some of them are based on factual observation, some are philosophical and logical, others are mathematical, and some are semantic. Sometimes the discoverer is led to a state of despair trying to solve the paradox whilst other times they result in entertaining puzzles that are used to confuse a philosophical opponent. The previously mentioned paradoxes of Zeno are a good example of an entertaining paradox in which a tortoise beats a professional runner in a running contest. We discussed this paradox in detail previously leading to the conclusion that something very special is happening in our everyday reality. Reality is fundamentally and structurally paradoxical. This paradoxical nature of reality is not incidental. The foundation of this paradoxical reality is the 'mother of all paradoxes', namely the paradox that we constantly refer to in this book. This is the paradox describing the simultaneous unity and multitude of reality. This paradox has its origin in the relationship between God and his energies, which is itself characterized by a simultaneous unity and multitude. This relationship is the elementary principle and forms the basis of all other paradoxes.

In essence, paradoxes can be divided into two categories: the first group of paradoxes refers to the relation, as described above, between unity and multitude and originates from the relation between God and his energies. These paradoxes refer to realities that exist and phenomena that can actually be observed. The paradoxes that have been discovered in quantum physics are a good example of this. The particle/field-of-energy paradox has been discussed in depth before. This paradox originated from observations that have produced the same results over and over again: particles sometimes manifest themselves as an energy wave and then again as a particle.

Apart from observations in laboratories of subatomic realities, there are also everyday examples of observations of the unity and multitude paradox, for example, the relationship between a substance and its properties. Substance and properties are inseparable and in that sense they are one. However, one single substance can possess several properties simultaneously. As a consequence, substance and properties are simultaneously one and different from each other. Take for example a familiar object such as a cube of white and sweet sugar. If we think more deeply about the relationship between the substance of the sugar cube and its properties, then we can conclude that this relation is simultaneously one and different. The substance of sugar is sweet, and the quality of sweetness is part of, and therefore identical with the underlying substance. We never perceive 'sweetness' apart from an underlying substance. The same holds for the property of whiteness related to the substance of sugar, we similarly never perceive 'whiteness' apart from an underlying substance. So, if sugar = sweet, and sugar = white, then this would imply that sweet = white. However, that is only partially true. Not all white things are sweet; salt is also white but not sweet. It is partially true, for the cube of sugar is simultaneously one and different from its properties. As such, the relationship between substance and properties is illogical, but yet we do perceive it. What is more, we eat it. Therefore the sugar cube, at least in terms of its relationship between its properties and its substance, is supra-logical.

At the core of this argument also lies the fact that the *Acintya* principle may only refer to existing things. The non-existing, the absolute nothing, cannot be a part of that, simply because it does not exist. In the following chapter, for example, we refer to the fact that *Sankara* and his followers used *Acintya* to justify the simultaneous existence and non-existence of illusion. The existence of illusion is used here by Sankara within the context of the absolute non-existence, more specifically the non-existence of the universe characterized by diversity and manyness. According to Sankara this universe of diversity and duality is an absolute illusion or Maya in Sanskrit. Opponents of Sankara would challenge his philosophy of absolute non-dualism by asking the question whether in his view the illusion is real and existed or whether it was unreal and did not exist. Paradoxically whether answered in the affirmative or in the negative, the outcome will be the same. If answered in the affirmative, i.e. the illusion is real, then it no longer is an illusion for it is real. If answered in the negative, i.e. the illusion is not real and does not exist, then it is not real either and does not exist. Sankara and his disciples would try to circumvent this paradox by stating that illusion is inconceivably real and not real simultaneously, effectively using the Acintya principle to justify irrationality, not based on experience but on semantic jugglery. It is the example of the square circle. On an even deeper level it fundamentally denies the validly of sense perception and empirical evidence. We directly experience the universe of diversity and manyness, and denying it is confirming it. Just as Descartes argued that denying or doubting our thought process is confirming it, and hence thought is the undeniable evidence for our existence, "e cogito ergo sum".

Other everyday examples of simultaneous oneness and difference are related to space and time, as expressed in the paradoxes of Zeno. Space and time are characterised by the fact that they simultaneously consist of an infinite number of small units - and in principle infinitely divisible units - while also being continuums. In other words, the distance from A to B can be described as an infinite collection of infinitely small points, or as a unity containing both A to B.

The simple three-dimensional expression of objects in space is also an everyday example of a simultaneous oneness and difference. A table has several sides, a top, an underside etc., but nevertheless it is one table. This is such an obvious point that we tend not to reflect on or question it. However, this simple reality of a table with a top, an underside and sides is a fundamental paradox. The reason that we do not perceive it as a paradox is because we observe it directly and probingly. If you were, however, to analyse the reality of this table in a few strictly logical steps; then you would end up with a paradox.

The second category of paradoxes, however, is mainly related to the definitions of linguistic terms and the logical constructions applied to them. Linguistic definitions and logical constructions can lead to contradictions in some cases. Often, a change of the definition can eliminate the paradox. A typical, well-known, and funny paradox is the liar's paradox:

"I am a liar; everything I say is a lie."

The paradoxical consequence of this statement is obvious: if what I say is true then according to the statement I am a liar and therefore the statement is not true. On the other hand, if what I say is not true, and I am not a liar, then I am in fact a liar. This type of paradox—and there are many of these paradoxes—primarily refers to the applied definitions of language use rather than referring to actual reality. No one is a liar all of the time, and there are probably few people who never lie. If the sentence were: "I am a liar and nearly everything I say is a lie", then the paradox would be solved.

This type of paradox also has to do with the phenomenon of self-reference, in which an object is part of the set to which it refers. Because of this self-reference, the paradox arises if the self-reference leads to a conflict with what is being claimed. In the scientific world the most well-known example of this is Russell's Paradox. Russell, a famous mathematician and philosopher, had just finished writing his book *The Principles of Mathematics* when he discovered that the fundamental principles of his book led to a paradox. His book was already finished, and in a hurry, he had to add an appendix to the book in which he explained that there was a problem with one of his propositions. This problem was later named after him. The paradox is the following:

"The set of all sets that are not a member of themselves is both a member of itself and not a member of itself."

This is a somewhat abstract and sometimes difficult to understand formulation, but it can be explained as follows. A set is a mathematical concept and refers to objects or things that are comparable or identical. A car, for example, is such an object of which can be said: 'the set of all cars.' So, each object that complies with the qualification of 'car' is, based on this definition, a member of this set. However, the set of all cars itself is not a car, so this set is not a member of itself. To continue this line of thought, we will use another example: the set of all things that are not cars. So, all things or objects that cannot be defined as a car, are thus a set. The set itself, in this case, is also not a car and is therefore a member of itself. Now, Russell concluded that if the set of all sets that are not a member of themselves is not a member of itself, then it becomes a member of itself. And vice versa, if the set of all sets that are not a member of themselves is a member of itself then that means that it must be a set that is not a member of itself. Hence, the paradox. Russell did not know what to do about this and it took him 10 years to find a solution for it. The solution, in the end, was that he defined the word 'set' differently, by stating that a set that is all-embracing excludes itself from such a set by definition.

Wittgenstein formulated the solution for this problem in a slightly different way. He stated that the logicians made the error by qualifying logical constructions as material things, as factual objects. By avoiding this, Russell's paradox completely disappears. Wittgenstein's analysis was correct. The interesting thing, however, is that we are again faced with the continuous interaction between unity and multitude: in this case a set, a unity, and the members of the set which are a multitude. In the end, all paradoxes can be traced back to this relation. Putting it slightly differently, the set is a whole, a unity. But in some instances the set, the whole, becomes a part of itself, and thus becomes a part among many other parts, which is a contradiction. The set is simultaneously the whole and part of the whole, it is one and many at the same time. Russell's paradox is a good example of this. The relation is a set, a unity, related to the members of the set, the multitude. This is a continuous area of tension. This also holds for the liar's paradox. This paradox is also faced with the fact that a statement is made about a set, a unity, namely the set of lies that are told by a particular person ("... everything I say is a lie") related to a multitude—the specific lies that are told by an individual. The paradox arises if the unity in turn becomes a member of the multitude and is actually one with and simultaneously different from it. Within the scope of logical constructions something cannot be one and different simultaneously, so the paradox has to be solved by extending the definition that an all-embracing set can never be member of itself, or refer to itself.

Another famous paradox is the 'barber's' paradox'. In a village there is a barber who shaves all people, and only those, that do not shave themselves. He advertises this on a large advertising board. The paradox arises when the question is asked whether or not the barber shaves himself. If he does not shave himself, then he does not shave all people that do not shave themselves. However, if he does shave himself then he cannot be the barber who only shaves people that don't shave themselves because he shaves himself. Whether or not this question is answered positively or negatively, in both cases it leads to a paradox. The first solution is to state the definitions and conditions differently. The hairdresser shaves everyone who does not shave himself, with the exception of the hairdresser himself. The other solution that basically comes down to the same thing, is stating that an all-embracing set cannot be a member of itself.

Some other important paradoxes which are well-known are related to wisdom, such as Hegel's paradox 'what we learn from history is that we do not learn from history.' Or Plato's remark that 'real knowledge is to know that we know nothing.' Or the hedonistic paradox: 'The more we strive for happiness or pleasure, the further away from it we get; the more we give up the desire for happiness and pleasure, the happier we become.' These paradoxes are more about use of language than they are actual and unbridgeable opposites. It can be substantiated that, in the case of Hegel's paradox, not learning is also a form of learning. So, this is about the definition and the use of the term 'learning'. In Plato's case we can state that knowing that you do not know is also a form of knowing. The hedonistic paradox refers to the definitions of happiness: for example, being free from desire, including of the desire for happiness can be defined as a form of happiness, possibly even a higher form of happiness.

A comparable proposition to Hegel's paradox is the proposition that is at the core of Vedanta doctrine and it is also paradoxical. This proposition is the following: "The purpose of our existence in this world is to realise that it does not have a purpose other than to make us realise that we should not be here." This is a metaphysical and philosophical proposition, but it nevertheless refers to a paradoxical reality. It is also comparable to the paradoxical and Zen-like statement: "The purpose is to not have a purpose." Despite the paradoxical nature of these kinds of statements, we know intuitively that they possess a deeper truth. The deeper truth is, as has been explained above, that concepts like 'happiness' and 'purpose' are not unambiguous concepts. Being purposeless can be a purpose in itself, but this would then be a purpose of a higher or different kind. Freedom from desiring a purpose or happiness can actually be a higher purpose and a higher form of happiness. It is the ambiguity of language that is the basis of these kinds of paradoxes. There are many comparable sorts of wisdom, such as the statement: "Everything that is good is bad and everything that is bad is good." What all these paradoxical statements have in common is that they are determined by definitions and the meaning of words, and the fact that the same words can have different meanings. The second category of paradoxes are, therefore, mainly puns and logical constructions that make use of the vagueness of language.

In summary, therefore, we can conclude that the first category of paradoxes are the most important, principally because they can be observed, and they objectively exist in the real world. The previously mentioned *Vedanta* principle of *Acintya bheda abheda tattva* refers primarily to this category of paradoxes. The paradoxes that have been discovered in quantum physics also belong to this category. The observation of sub-atomic particles that manifest themselves simultaneously as a particle and a field of energy are factual observations that refer to an objective reality. These are not logical exercises of thought, mental constructions, or linguistic ambiguities. These paradoxes namely refer to observed facts and events that cannot be explained within the scope of existing laws of nature and based on the common laws of logic. They relate to physical realities that are observed, but which contradict the common laws of nature and the laws of logic.

Such observations however, transcend logic, because logic is in the end also a derivative of observation. The three laws of logic—the law of identity, the law of contradiction, and the law of excluded middle—are based purely on observation and intuition. The fact that something is what it is, is so because we observe it that way. The fact that an apple is not a pear is an unambiguous observation and empirical conclusion. The fact that all things, whether apples or pears, either exist or do not exist, is fundamental to the way we experience reality. There is no third alternative that sits in between existence and non-existence. These issues, which were thoroughly discussed in previous chapters regarding the laws of logic, underline the limitations of our intellectual and logical capacities.

Another important comment regarding the *Acintya* principle is that it is characterised as supra-rational and not as irrational. The crucial difference is exactly what has already been substantiated above, namely the difference between objectively existing and non-existing things. Something that exists objectively but which is paradoxical is supra-rational since it is empirically true even though it entails a logical contradiction. On the other hand, something that is irrational is both empirically and logically non-existing. A square circle is a good example to illustrate this. A square circle is irrational: it is logically impossible, and it does not exist empirically. No one has ever seen a square circle. Factually, a square circle is a contradiction that exists because of the wrong use of language. A circle is round by definition and never square. If we all would decide and agree to call a circle a square and a square a circle, then this is just a way of altering linguistic definitions and agreements. The properties of the circle itself and the square do not change because of this. Many of the paradoxes in the second category, as we have discussed them so far, are part of this category regarding the use or misuse of linguistic definitions. It is important to highlight this point, for it denies us the ability to use irrationality as a justification for presenting an irrational world view. The existence of the supra-rational reality must never lead to a termination of the rules of logic and an opening of the door to philosophical and scientific chaos.

The Key Paradoxes That Define Our Objective Reality

We now come to the main point of this chapter in which we describe the key paradoxes that make up our objective reality and how these paradoxes relate to and actually describe the identity and nature of God. Below is an enumeration of the various paradoxes as they manifest themselves in reality. This enumeration overlaps with the description of the paradoxes as they manifest themselves in consciousness as described in Chapter 10.

1. Simultaneous oneness and multitude

The principle of simultaneous unity and multitude has a direct link with the fundamentals of the New Physics. We have previously referred to the fact that quantum physics has concluded that particles behave at a sub-atomic level as both particles and fields of energy. Furthermore, it has been established that both of these states of being are interchangeable and that these characteristics of sub-atomic particles do not fit within the normal rules of logic and the normal laws of physics. In quantum theory this is called the particle-wave duality. In 1927 the astrophysicist A.S. Eddington formulated that the paradoxical behaviour of particles and waves could be called wavicles. He said:

"We can hardly describe such an entity as a particle or a wave and as a compromise we might best describe it as a wavicle." [89]

This term was later popularized by various physicist and mathematicians. Capra talks about this in his book *The Tao of Physics*:

"Time and time again when the physicist asked a question in an atomic experiment, nature answered this question with a paradox, and the more we tried to get things clearer, the sharper the paradoxes became. It has taken us a long period of time to accept that these paradoxes are part of the intrinsic nature of the core of physics and to realise that these [paradoxes] present themselves over and over again when an attempt is made to describe atomic events in terms of traditional physics." [90]

A few paragraphs later Capra states:

"The sub-atomic unities of matter are very abstract entities that have an aspect of duality in them. Depending on how they are observed they sometimes present themselves as particles and sometimes as waves." [91]

The conclusion is inevitable that the fundamental building blocks of matter are paradoxical, which exceeds the capacity of our minds. We know that it is true, since it is observed, but at a fundamental level these characteristics of matter and energy are supra-logical.

This paradox has other far-reaching consequences. Because of the field-particle paradox, space and time should be placed in an entirely different perspective. We mentioned previously that in a field all parts of the field are connected to each other. Because of this, space and time can be eliminated. Each event within a field has a simultaneous effect on every other event within that field. In the chapter on the extraordinary nature of reality, we referred to the quantum phenomenon of entanglement where Einstein talked about 'spooky' action at a distance. It appears supernatural because for an object outside the field there is indeed a difference between one part of the field and the other part of the field. Therefore, an observer will experience this elimination of the limits of space and time as supernatural.

2. Simultaneous 'fullness' and 'emptiness'

In the world of mass, movement can only occur where there is 'empty' space for an object to move into. Something that is completely full would prevent any movement. At least this appears to be the case in the world of mass and matter. In the energetic universe, the relation between space and energy works differently. In its pure energetic form energy is not composed of particles and it does not have mass. In that state energy does not differ from the space that it is in. What is more, energy *is* space and space *is* energy. This implies that actual empty space does not exist, but that space is a function of pure being. Space thus understood and thus defined is infinitely flexible and despite its fullness is able to accommodate motion. Space, as a function of being is starkly contrasted by the materialistic vision of the universe, where space is regarded as a function of non-being: the absolute void, pure nothingness. This is not the first instance where the materialistic view and the theistic view are diametrically opposed. In an earlier chapter it was concluded that the theistic view is based on the notion that the origin of everything is infinitely full, while the atheistic view states that the origin of everything is pure nothingness. Energy is full and simultaneously it enables motion; it is unchanging and permanent, while it continuously changes and moves. Furthermore, energy is the cause of movement and change, but it is also that which changes and moves.

The descriptions of dark energy similarly imply that space is not empty, but that space itself is energetic. The acceleration of the expansion of the universe is attributed to the existence of this dark energy. While energy in relation to mass and matter manifests itself as gravity and attraction, energy also manifests itself as movement and expansion. In this way it creates space as a function of its expansion. In the end the relation between space, motion and the motion of individual objects in space, is also a function of the Acintya principle simultaneous oneness and difference, space representing the oneness principle and individual objects representing the manyness principle, that co-exist simultaneously.

3. Substance and qualities are simultaneously one and different

The relation between substance and its qualities is also determined by the oneness-difference paradox. Philosophers have been debating the relation between substance and its qualities for thousands of years. According to Plato they were different; according to the nominalists—a philosophical movement—they are one. At the level of the material world this is an unsolvable paradox. After all, if they are one, then how can it be explained that a substance possesses several properties? A cube of sugar is white and sweet. If these properties would be completely one with the underlying substance, then all white things would have to be sweet and all sweet things would have to be white and this is not the case.

On the other hand, we never perceive sweetness or whiteness without them being linked to a certain substance. According to Plato these qualities exist at a higher, universal level and they are projected, as it were, onto various objects. We have never witnessed these 'universals', and therefore this is nothing more than a hypothesis. What we do perceive are specific objects made of substance and properties.

What we also perceive and are able to determine is that energy possesses all actual and potential properties (described in point 4). Based on that understanding, the theory of Plato regarding the existence of universals can be seen in a different perspective, namely that energy is the infinite reservoir of actual and potential properties. In this sense, Plato's concept of the projection of qualities refers to the qualities within the energy state transforming themselves from potential into actual. Nevertheless, the relation between substance and properties is a relation of simultaneous oneness and difference.

4. Simultaneous existence of actual and potential qualities

There is another implication with respect to the relation of simultaneous oneness and difference between qualities and their underlying substances. Since a substance can contain many different and even opposing qualities simultaneously, on a deeper energetic level it implies that energy is infinitely full and possesses all actual and potential qualities simultaneously without them 'pushing each other aside'. Energy represents the infinite field of the unlimited possibilities in which all actual and potential qualities are present. This explains why we sometimes witness the creation of new qualities, for example, in the process of chemical reactions. The simultaneous presence of all potential and actual qualities can be compared to discoveries in quantum theory where sub-atomic reality is described as a reality that is undetermined, and which is filled with potential possibilities. The potential possibilities are transformed into an actual reality the moment that an observation takes place. We discussed this in a previous chapter as part of the 'cat paradox'. In quantum jargon this concept is called the collapse of the wave function whereby the collapse refers to an undetermined field of energy which is suddenly transformed into an actual and determined object. The crucial point is that quantum theory fundamentally assumes that reality at its deepest level consists of a state of energy that is filled with all potential and actual qualities.

In point 2 above, we also established that energy is full and simultaneously moving; it is unchangeable and permanent while it continuously changes and moves. Energy is the cause of movement and change, but it is also that which moves and changes. These are all incredibly difficult concepts to comprehend and emphasise that energy possesses extraordinary characteristics. These characteristics can be observed, despite the fact that they are paradoxical from a logical point of view. For this reason, these paradoxes can be qualified as supra-rational, contrary to irrational. The essence of a supra-rational paradox is that it can be observed. This observation exceeds all logical considerations and limitations.

5. Simultaneous change and permanence

In chapter 10, we discussed the phenomenon of 'consciousness' and one its unique characteristics: the ability to experience and undergo observations and sensations that are by their very nature constantly changing, without itself changing. This paradox also holds for the world of energy and mass where we experience the concept of Being, or permanence, in a reality that is constantly changing. Indeed, the Greek philosophers Parmenides and Heraclitus are well-known for having held completely opposing views from one another on this subject. Parmenides (about 515 BC), Zeno's teacher, supported the idea that only permanent Being is real, whereas change is simply an illusion. On the other hand, Heraclitus (540-480 BC), a contemporary of Parmenides, promoted the exact opposite idea. According to Heraclitus everything was subject to a perpetual change. The only thing that was permanent in this process of perpetual change was change itself.

Whilst Parmenides had a fanatical follower (namely Zeno), Heraclitus also had a fanatical follower, named Cratyclus. According to Cratyclus, one would never be able to step in the same river, for the moment one steps in the river, the river has already changed. Even having a conversation was pointless, according to Cratyclus, for during the conversation the speaker, the words, the meaning of the words, and the listener would all change. When someone talked to him he would move his finger as a sign that he had understood the message, but he would not answer. In his view there was no longer any point in responding as all the facts and circumstances had already changed.

However, we can observe that reality contains both elements, change and permanence simultaneously, whereby one cannot exist without the other. In this instance permanence represents the principle of oneness, whereas change represents the principle of manyness and diversity. There is a hierarchy though. Change only has meaning in relation to something that does not change. It is not the change that changes, it is SOMETHING that changes and this 'something' implies an underlying permanence. 'Being' is the foundation of permanence, and change is a function of being. This is also the view from Vedanta doctrine and leads us to the conclusion that change and permanence are both real in a paradoxical sense.

6. Space and time characterize themselves as a simultaneous unity and multitude

The famous paradoxes of Zeno that we referred to previously are a mirror of the principle of simultaneous unity and multitude. In this case the paradox refers to the characteristics of space and time in which unity and multitude exist simultaneously. Both space and time characterize themselves by the fact that they are simultaneously infinitely divisible and indivisible. This simultaneous unity and multitude is so intertwined with our daily lives that we hardly think about it. However, movement would not be possible without this principle. For example, we are able to move from point A to point B because A and B are two different locations that are nevertheless connected to each other. This connection makes it possible for me to move from A to B, since A and B exist in the same reality, the same space. Again, this is such a matter-of-fact that we do not think much about it. If we reflect deeper, however, it is actually quite perplexing.

Zeno took the point of view that space only exists of multitude or, to be more exact, an infinite multitude of infinitely small points. Because of this infinity, it was impossible to even bridge the smallest of distances. What is more, according to Zeno and his teacher Parmenides, movement was impossible and everything we perceive as movement is only an illusion. The reality is that movement is possible because the multitude of points in a space are linked together by the unity of space itself.

7. Unity and multitude: the fundamental principle of reality

Factually, this principle is the most important and paradoxical cornerstone of reality. The principle of simultaneous unity in multitude and multitude in unity makes reality unfathomable. Indeed, it is the very basis of the mystery of existence as we have just discussed and demonstrated above. We now summarize all these points, which together with our conclusions from the teleological and anthropic principles, can be combined to form a universal definition of what God really is, a definition that includes a thorough understanding of our contemporary scientific reality.

Energy is not without attributes

Energy is the foundation and base of the universe, and mass and matter are a transformation of energy. While energy exists in the form of mass (or matter), it also exists in a completely different way: purely energetic, without mass, and non-material. This purely energetic condition of being does not imply that energy is shapeless and without attributes. On the contrary, energy is the reservoir of all actual and potential attributes. This means that the energetic, non-material universe is characterized by attributes and characteristics, comparable to the attributes that we know in this universe, but these attributes are of a different quality and a different order.

This unity-in-multitude is infinitely intelligent

The infinite energy that forms the foundation of the unity and multitude of the universe is infinitely complex and infinitely intelligent. This intelligence and complexity manifest themselves abundantly in the physical universe, as articulated and explained in the teleological argument, the anthropic principle and the principle of cosmic fine-tuning. This intelligence is pure consciousness, and it is the all-determining driving and creative force behind the universe. Consciousness itself constitutes the foundation of the principle of simultaneous unity and multitude, for unity in multitude is the very essence of consciousness. Conscious intelligence entails the ability to perceive, analyse, understand, and organise the multitude of objects from a centre. Consciousness *IS* the unity and the multitude; the unity is the individual identity, the personality, the centre. The individual gathers information by means of his consciousness and integrates this information into an experience. This experience is a synthesis in space and time of a multitude of sensations and information, whereby the multitude is integrated into a unity, or what we call 'understanding'. The multitude is therefore consciousness that is aware of itself, of its surroundings and that absorbs the multitude of observations and sensations. This multitude of information is accordingly gathered and concentrated in the centre, the individual. This is such a remarkable process, but at the same time such a matter-of-fact that we are often not even aware of it. It is like looking at the wall in front of me, while unable to see my own eyes. Or two fish swimming in the ocean and one fish remarking to the other fish how much water there is, and the other fish responding: "what do you mean I do not see any water."

Moreover, consciousness is active. The process of observation and understanding represents the passive function of consciousness. Will power and desire represent the active function of consciousness. Therefore, consciousness not only has understanding and intelligence at its disposal, but it can also employ these attributes in the process of ingenious creation.

This process of creation takes place based on pure willpower. Willpower does not have any other source than itself; it does not have a cause, but it is the cause of all other causes. For this reason, the creation of the material universe can only occur based on consciousness, intelligence and willpower. Any other explanation is ultimately insufficient.

This conscious and intelligent unity is God

All attributes and qualifications, as were discussed in this chapter, come together in an entity that can only be qualified as God. God exists and is a tangible reality, he is the fundamental, intelligent, eternal and infinite unity of the universe from whom flows pure consciousness and pure energy. In the following chapter the nature and attributes of God will be discussed in more detail, based on the scientific and philosophical conclusions reached thus far, and based on the insights of the Vedanta philosophy.

"And I am the basis of the impersonal Brahman, which is immortal, imperishable and eternal and is the constitutional position of ultimate happiness."

Krishna, Bhagavad-gita Ch. 14.27

"Although God is situated in His own abode in the spiritual sky, His energy is distributed everywhere."

A.C. Bhaktivedanta Swami Prabhupada, Journey of Self-Discovery, Chapter 2

Personal or impersonal?

Based on all the evidence, arguments and discussions presented so far, we have come to the logical conclusion that God must indeed exist. The next logical step is to try to better understand who or what God is. In previous chapters, it has already been established that God is intelligent and conscious; that he is eternal and infinite; that he can only be good by definition; and that he is simultaneously one with and different from his energies in a supra-rational manner. Within the scope of these characteristics and qualifications there are two fundamental options that we have not yet discussed and answered in detail: is God personal or is he impersonal? Is he an impersonal state of formless being made of pure energy without centre, or is his being expressed in a personal identity with a centre and endowed with attributes, shape and personal qualifications.

In the first chapter, where, amongst other things, the definitions of the concept of God were discussed, this question was addressed very briefly. Theistic philosophies were divided into monism and monotheism. The monist version regards God as impersonal, as an infinite field of energy of pure consciousness without centre. Monotheism, on the other hand, also characterizes God as infinite and all-pervasive, but at the same time as personal and transcendental. This implies that he is separate from his energies and possesses a personal identity. Of course, these are very difficult concepts and abstractions if we project them on the infinite reality that is God, a reality that exceeds our understanding in almost every way. Nevertheless, we can pose these questions, and despite our intellectual and psychological limitations, they are very meaningful questions.

Sankaracarya

The question is, furthermore, completely logical. Everything we know in the universe can be divided into two categories: personal consciousness on one hand and impersonal, not-conscious energies on the other hand.

As stated, an important conclusion from previous chapters is that God is conscious and intelligent. Everything we know and understand about consciousness has, of course, something to do with our own consciousness. We experience our consciousness as strictly personal and as a part of our personal identity. From this perspective—and drawing from our own experiences—it is logical that God, as a conscious being, is also personal. The Vedas agree with this conclusion. Even though there are important movements within the Vedic tradition that describe God as impersonal, this does not match the origin of the Vedic insights. The rise of the impersonal interpretation of the Vedas only came into being in the 7th century AD, with the advent of the great Indian philosopher and mystic Sankaracarya. Sankara gave an explicit, impersonal and monistic interpretation to the Vedanta-sutra, the philosophical core of the Vedic doctrine. This interpretation is also called 'Advaita Vedanta'. *Advaita* literarily means non-duality, contrary to '*dvaita*', which means duality. Advaita Vedanta is based on monism, a well-known and influential philosophical movement in Western society and in India. Monism states that our universe, with its multitude of objects and attributes, is only an illusionary appearance of an underlying, all-pervasive, undifferentiated and formless oneness. This unitary oneness is real, and the world of phenomena and the multitude of objects and attributes is false. This underlying oneness exists of pure energy and, according to Sankara, this state of energy consists of pure consciousness. Sankara calls this infinite, undifferentiated ocean of pure energy *Brahman* and equates this to an impersonal God.

Prior to Sankara, however, the Vedic texts and conclusions were of a monotheistic nature, and they were interpreted as such by the various traditions. Sankara's impersonal interpretation led to fierce philosophical debates within the various Indian Vedanta schools. The core of these debates focused on three matters:

1. **Whether God is confined to mere oneness without attributes and without shape, and therefore impersonal, or whether he is personal, possessed of shape and attributes;**

2. **Whether the material universe is false, an illusion, or**

whether it is a real — although temporary —and an emanation from, and transformation of God;

3. Whether we, as living beings, are both qualitatively and quantitatively one with God and therefore in the end God himself, or whether we are only qualitatively one with God, but quantitatively different;

This eclectic discussion has had far-reaching consequences. Since the impersonal vision of reality implies that reality is an undifferentiated and absolute oneness, all duality and diversity must therefore be an illusion. The material world which is full of diversity and multitude was therefore cast aside by the impersonalist philosophers as an illusion (in the meaning of something that does not exist and is not real). These philosophers and mystics were therefore called *mayavadis*, which means 'those who characterize the world as an illusion'. Furthermore, they stated that each living being is not just a part of God, but that after liberation from illusion each living being literally becomes one with God. In other words, we become God in every way, totally and completely without distinction. In fact, we are already God, but we just do not know it yet.

The personalists have an entirely different vision of God, his energies, and the position of living beings such as ourselves. They define God as personal, who manifests and emanates his energies from his personal centre. These energies are both personal and impersonal, and therefore God possesses a personal and an impersonal aspect. The core and centre, however, is his personal manifestation and his personal identity. His energies form the creation, and they consist of the same qualitative substance. Yet, they are also different from him in the same way that the sun is simultaneously one with and different from the sunlight it radiates. If I enjoy the pleasantly warm sunlight on a sunny day, this is not the same as if I were to face the power of the entire sun. Therefore, sunlight is qualitatively equal to the sun and it originates from the sun, but quantitatively, there is a massive difference.

Sri Caitanya Mahaprabhu (1486-1534). Caitanya was the founder of the Vedanta school that united all different Vedanta movements. His school, based on the principle of 'acintya bheda abheda tattva', united the monism of Sankara with the dualism of various theistic traditions. Caitanya stated that God is simultaneously one with and different from His energies in a supra-rational way. This approach was revolutionary and innovative and made way for a new metaphysical approach. In this approach reality consists of a fundamental paradox that manifests itself in all aspects of existence.

Consequently, personalists define the relation between God and His energies as simultaneously one and different. The previously mentioned influential Indian mystic and philosopher Caitanya Mahaprabhu defined this relation as *acintya bheda abheda tattva*, which means that God is, in a supra-rational inconceivable way, simultaneously one with and different from his energies. This philosophy is the cornerstone of the personalist, theistic Vedanta school. This school furthermore teaches that the living being may be qualitatively one with God, but quantitatively always different. A living being can never become God within the full meaning of the word. He is a part of God and will always be a part of God, and within that sense he is constitutionally subordinate to God. It is a hierarchical relationship that a small number of the infinite number of living beings have some difficulty in accepting. The hierarchical relationship is therefore also the foundation for our separation from God and our descent into the material world (as has been explained in previous chapters).

In summary, personalists maintain that God is personal and that his personality represents the ultimate and non-reducible reality. They regard personality as the ultimate explanatory principle and not just as an incidental attribute of reality—on the contrary, reality is an attribute of personality. In a similar way personality is not just an attribute of God, God *is* a person. Personalists have three important arguments in support of a personal God. Firstly, both personalists and impersonalists agree that God is conscious. Consciousness, however, is inconceivable without personality. The core and definition of consciousness is the conscious observation, the experience, based on the object-subject relationship: a subject, or an I that observes, and an object, or a non-I. Ramanuja confirms this and he also states that observation consists essentially of three components: the conscious observer, the process of observation itself, and the object of the observation. Therefore, the conscious observer is by definition personal. An impersonal consciousness, a consciousness without identity and an I, is inconceivable and impossible. It is like the matter of a square circle – a contradiction in itself.

Secondly, the universe is filled with personalities and filled with attributes. We are personalities, we possess attributes, and the universe is filled with attributes. Based on the causality principle these attributes have to be present in the origin of the universe, in God. However, in order to avoid the causality argument, impersonalists state that the universe, filled with attributes and personalities, is actually just an illusion. If the universe with all its attributes and personalities simply does not exist, then the cause—God—can easily be impersonal, shapeless, and without attributes, without violating the causality argument. Whilst denying the existence of the universe seems a very far-fetched argument to make, it is nonetheless, the core of Sankara's doctrine. Therefore, the third argument of the personalists focuses on the fact that this denial of the existence of the universe seems to be impossible, both logically and empirically. It is important to note at this point what exactly we mean by illusion. Sankara's conclusion that the universe, the world of duality, forms and attributes must be an illusion does not refer to the meaning of illusion as we described in chapter 7. There, we also referred to *Maya*, or illusion, but within the meaning of an incorrect interpretation of reality. Ramanuja uses the example of a piece of rope that is mistaken for a snake in the dark. The piece of rope is real and exists, but the observer interprets this differently because of ignorance. However, Sankara's interpretation of this example is completely different; his explanation is that neither the rope nor the snake actually exist, and the startled observer is merely seeing an hallucination — an hallucination generated by his own consciousness that is accordingly projected outwards onto the screen of the ocean of pure, undifferentiated energy. According to Sankara the universe does not exist at all and is just a mental projection of the conscious individual. Factually, Sankara thus states the proposition of subjective idealism, the proposition that the individual observer creates reality.

Whilst this is not the formal position of Sankara, some of his followers do agree to this philosophical position even though subjective idealism, as has been explained before, is totally indefensible. Sankara's position is more implicit than explicit. Sankara states that the illusion of the universe is created by an objective, cosmic power that the individual observers are subjected to. How, why, and what the status is of this illusion generating cosmic power, or whether this cosmic power itself is an illusion, remains a complete mystery in the teachings of

Sri Ramanujacarya

Sankara though. Therefore, it is exactly at this point that he is attacked by his philosophical opponents.

Ramanuja, a fierce opponent and critic of Sankara, argues in the following manner. If the world is an illusion, then the following question can be asked: is this illusion real and does it really exist, or is it unreal and does it not exist? In other words: Does the illusion exist, or does it not? If the illusion is real and therefore exists, then the world, including the attributes and duality that are part of the illusion, also exists. If the illusion is unreal (the illusion does not exist), then this means that the world is not an illusion, and therefore exists. Denial or confirmation both lead to the same result, namely that the universe inclusive of its form, attributes and specific objects exists and that personality exists.

One of the cornerstones of Sankara's Advaita Vedanta is the notion that God is in every sense incomprehensible and indescribable. This incomprehensibility and indescribability are also applicable to its illusionary energy, *Maya*. Followers of Sankara therefore defended their position against the criticism of Ramanuja by referring to the fact that *Maya*, illusion, is incomprehensible. The core of this counterargument is that illusion exists and does not exist simultaneously in an incomprehensible way. This appears to be similar to Caitanya's doctrine of *Acintya bheda abheda tattva*, with the crucial difference that Caitanya's doctrine only refers to existing realities. The simultaneous oneness and manyness of objects, which are the two sides of the Acintya equation, refer to existing empirically observable realities. This does not hold for Sankara's version of *Acintya*, because one side of the equation refers to the absolute non-existence of something, which makes the equation invalid. As we have established previously, the absolute non-existence does not exist by definition, and therefore existence and absolute non-existence cannot co-exist simultaneously. Even the *Acintya* principle cannot eliminate this impossibility.

The supra-rational and mysterious nature of God

The supralogical paradox

The importance of the *acintya bheda abheda tattva*-principle, the supra-rational paradox, has been thoroughly discussed in previous chapters. We have repeatedly referred to the simultaneous unity and multitude of reality throughout this book. According to the *Acintya* doctrine, this is the cornerstone and the core of the incomprehensible and supra-rational nature of God. This supra-rational nature of God manifests itself in three ways. Firstly, there is the internal nature of God, his being of pure consciousness. Consciousness consists—as we have described previously—of an identity, which is the centre, or unity, that is able to manifest itself simultaneously as a multitude. Secondly, God expands by means of his energies, his emanations. These energies represent his external nature or his non-self and they are simultaneously one with and different from God. Thirdly, the *Acintya* principle of simultaneous unity and multitude manifests itself within the manifestations and expansions of his energies. This means that the supra-rational paradox manifests itself within his energies, including the material universe. In chapters 9 and 12 we have given a thorough and detailed description of this. The *Acintya* principle is the core of the mystery of existence, since it is fundamentally paradoxical. The meaning of a paradox is that its existence is impossible according to logic, but that it nevertheless exists based on empirical observation. We perceive something that is rationally inexplicable. However, since we perceive it—and in this respect perception is superior to logic—it nevertheless exists, and we have to conclude that reality is incomprehensible and supra-logical, or *Acintya*.

Infinity

The *Acintya*-principle is not just determined by the paradox of simultaneous unity and multitude of God and his energies. The *Acintya*-principle is also determined by the infinite nature of God and the infinity of his energies. God is infinite and eternal, and therefore he exceeds our comprehension in this way too. We are, as living beings that are part of God, eternal in time, but finite in space. As finite beings it is by definition impossible to fully understand infinity. We can however understand infinity in a qualitative way, in a sense that a part is a qualitative representation of the whole, just like a drop of water from the ocean represents a (partial) mirroring of the chemical composition of the ocean. The drop can be more or less contaminated than the rest of the ocean, but the basic chemical composition H^2O is the same everywhere. An even better example is the example of a hologram in which each part of the hologram contains the image of the whole. Reality is holographically constructed and that provides us as finite beings with a peek beyond the veil of what separates the finite from the infinite.

God is infinite and his expanded energies are similarly infinite. In philosophy and mathematics a distinction is made between actual and potential infinity. Space, for example, is actually and currently infinite, which means that the infinity of space already exists. A series of numbers, on the other hand, is potentially infinite. Whatever number you chose, you can always find a bigger number. God is both actually and currently infinite, but he is also and simultaneously potentially infinite, which enables him to keep on expanding his infinity into further infinity. For this reason, God is able to create his so-called plenary expansions, as described in the Veda's, which are complete duplicates of himself including all of his attributes and characteristics, and which are as infinite as he himself is. Nor is God reduced or changed in the slightest by such expansion. Infinity minus infinity remains infinite and infinity minus an infinity multiplied by infinity is still infinite. Therefore, it is not surprising that we have difficulty understanding God. He is *Acintya*, supra-rational, paradoxical and He is infinite, the comprehension of which is a challenge even for the most brilliant mind.

The nature and the being of God

Earlier in the chapter we have ascertained that God is conscious and, therefore, he must have a personal identity. This means that he possesses attributes that we also possess as persons. For example, he has a form or body with attributes such as a shape, colour, size, and specific features as well as a personal taste, preferences, and emotions. However, God's form, attributes and characteristics are of a superior spiritual, transcendental nature. These are distinctly different from the characteristics of personalities residing in the material world, possessed of temporary material bodies that are made of an inferior energy, matter. We will discuss the specific transcendental features of God as a person in more detail in the following chapter. For the remainder of this chapter, we will explore how the essence of God is defined in the Vedas and what this means for understanding our own essential being.

The Vedas describe the essential nature of God as *sat-chit-ananda*, a Sanskrit term which means eternal pure existence, pure consciousness and knowledge, and pure bliss. *Sat means "pure existence" and* refers to God's characteristic of eternal existence, and the eternal existence of his emanations and energies. These energies accordingly transform into the infinite diversity of manifestations and creations, both materially and spiritually. All these manifestations are emanations and transformation of God and his Divine energies, and all of them are fundamentally eternal.

Chit, the second characteristic, refers to pure consciousness and intelligence that manifests itself as all pervasive knowledge and an omnipresent consciousness. *Chit* is the intelligent force behind the universe, the creator, the organiser and the maintainer of the infinite complexity of the universe. Chit is also the omnipresent consciousness of God that permeates each atom, each cell and each fibre of creation. While this sounds truly extraordinary, we actually have a similar experience of this characteristic with regard to our own consciousness, albeit at a very limited scale. We, as human beings, also possess a consciousness that permeates each part of our body and reaches out beyond the limitations of our body by means of our senses and the media employed by our senses, such as light and sound. This awareness is also truly remarkable if we really think about it. Based on this awareness it is not impossible to extrapolate our own experience towards a situation in which these attributes of awareness occur at an infinitely much larger scale.

The third attribute is *Ananda*, the emotional condition of unlimited bliss. This last attribute represents the ultimate purpose and the culmination of the previous two attributes, namely existence and intelligence. Existence and intelligence are meaningless if they don't culminate in bliss or happiness. The experience of bliss is the ultimate purpose of existence, something we all recognize as living and sentient creatures. The fundamental motivation of each person and of each conscious being is driven by our desire to experience happiness and to avoid pain. The reason we have this innate desire for happiness is that as a part of God we are also sat chit ananda and made of the same qualitative substance and attributes, except that we possess these attributes in a quantitatively limited amount. Since we have the same attributes, it is our elementary and natural condition to fully experience these attributes. In our present condition however, we are disconnected to some extent and prevented from experiencing these attributes in full.

Another interesting feature of God is that since God is infinite and perfect and since he is in a permanent state of bliss, he does not have any tasks or missions that have to be performed. He does not need to 'work' as we understand this concept when we think of our own lives. His creative ability does not require effort; the thought or the desire is reality and reality becomes immediately manifested. In such a situation we may ask ourselves whether or not this would lead to boredom and lack of purpose, especially considering the fact that this situation lasts forever. However, the notion of boredom is related to our own personal emotional limitations. We cannot imagine, in our current conditioned, limited psychological state, a permanent state of emotional fulfilment and happiness. This is because emotions and sensations cannot be 'imagined'; we can only experience them. For example, it is impossible to really understand the feeling of falling in love with someone until you actually experience falling in love. So, in this way our emotional understanding is largely limited to the emotional experiences we have had.

A second issue is that our emotional experiences are characterised by duality or opposites, such as happiness versus sadness, and when we are in any given emotional state it is very hard to feel the opposite emotion, and in extreme cases, to even doubt whether it is possible to feel such an emotion. For example, when we are feeling extremely unhappy, it is difficult to remember happy moments. We can remember that we were happy, but it is nevertheless very hard to re-feel that same feeling. Another more everyday example illustrates what is meant here. Imagine being very hungry and really looking forward to having a palatable meal. Now when we are hungry we often tend to put a lot of food on our plate. The overriding feeling is that we will be able to eat a lot of food because we are so hungry. However, after dinner, having probably eaten too much, the feeling of hunger is replaced by a feeling of being fully satisfied. At that moment we are incapable of remembering the feeling of being so hungry, and even worse, when we think of food we get nauseous. Both the experience of being hungry and the experience of being satisfied display a sense of absoluteness, at least at that moment. With this we mean that when you are experiencing being hungry you cannot imagine what it feels like to be satisfied and vice versa. Therefore, emotions and sensations seem to be of an absolute nature at the moment that they are being experienced; they are difficult to put into perspective. For this reason, a depressed person, in his state of depression cannot imagine that he or she will ever feel better, which is in fact one of the symptoms of a depression and perpetuates the depression. The fact that a negative emotion can have such an absolute and permanent nature enhances that emotion and can lead one into a negative, downwards spiral.

On the other hand, positive and blissful emotions also seem to be absolute at the moment they are being experienced and they lead to a positive spiral of positive moods that enhance each other. In such a mood one is not bored, and time is characterised by the absence of it. It is a familiar phenomenon that when people have a good time, time flies. Boredom has everything to do with our experience of time, where the absence of time eliminates the feeling of boredom. On the other hand, as one's mood becomes more negative, time slows down and the minutes pass by so slowly. We all know this experience as well.

Emotional limitation, expanded consciousness

The reference to emotional limitation demands more explanation. This limitation has to do with the fact that many people cannot imagine that we are able to experience a permanent, even eternal state of happiness. When we think of eternal happiness in a heavenly settings, images come to mind of beautiful angels sitting on white clouds forever playing on harps. While attractive and serenely peaceful it also conveys a sense of boredom and lack of excitement. It looks nice for a while, but to spend the rest of eternity in this manner? This example is of course a banal oversimplification of commonly held views, clichés and folksy tales, but they are revealing nonetheless. These are still the sorts of images that people often associate with spiritual happiness. There are many jokes about the fact that hell may be a terrible, but that at least it is an exciting place where lots of things are happening! These perceptions and its imagery have everything to do with our emotional limitation.

We are parts of God and as such we are connected to God at the deepest possible level, the level of consciousness and vitality. To try to visualize this connection the comparison of a light bulb that is connected to a power station by means of a cable, is an apt comparison. In this example, we are the light bulb and God is the power station. As a light bulb we have the potential to emanate light, but we depend on the electrical current that, via the power station, reaches the light bulb. The power station is always on, constantly generating electricity, but the cable connecting the power station and the light bulb is equipped with a dimmer type switch. The dimmer can be used to increase or decrease the flow of electricity. In this example the dimmer can be compared to our ego and as our ego grows, the supply of electricity is dimmed proportionately. We operate the dimmer, indirectly, by means of our desires and subsequent actions. The more our desires are focused on our ego and the satisfaction of egotistical desires, the more the dimmer is activated. As the dimmer increasingly blocks the flow of electricity, the luminosity of the light bulb diminishes until there is only a faint glow. That faint glow can be compared to our material consciousness, dimmed and ignorant, with a limited amount of energy and awareness.

However, as we adapt our desires and actions we begin to turn the dimmer the other way allowing the current of electricity to start flowing again, and consequently, our consciousness expands. The broadening and expansion of our consciousness has both an intellectual and an emotional dimension. Intellectually, we experience the expansion of consciousness as enlightenment, as an increase in the clarity of our thinking. It also gives us a higher intuition and our understanding is sharper. Emotionally, we experience the expansion of consciousness as an increasing sense of joy, happiness and fulfilment. Because of the absolute and overwhelming nature of emotions this has a powerful effect on our psyche.

As we stated above, in our current state of mind, we cannot imagine a permanent state of happiness without that happiness decreasing in intensity at some point. Our experiences of happiness are often characterized by high points interchanged with low points. The fundamental reason for this is that our experiences of enjoyment are linked to sensory experience and sensory joys, in other words they are linked to our interaction with matter. These are, however, temporary and ever decreasing in intensity due to the temporary nature and limitations of matter. At the most fundamental level, it is impossible for the soul, which is constitutionally made of pure spiritual energy and consciousness, to fulfil its emotional needs through any amount of exploitation of matter. Spiritual happiness does not depend on external stimulants as it is generated from the centre of consciousness. When we connect to this centre of consciousness we simultaneously connect to our source of consciousness, God. As a result we will undergo the intense experience brought about by the increase in the supply of spiritual energy and the subsequent expansion of our consciousness.

Thus, as our emotions reach deeper layers of happiness, we become aware of the fact that the depth and intensity of our emotions are unlimited. To our astonishment we realise that we can experience feelings that we did not know exist. Basically, a whole new universe opens up for us of with an array of emotions, sensations, colours, smells, tastes, sounds and feelings of an unknown ecstatic intensity, filled with unlimited variety and a never-ending sense of delightful amazement. In this state of consciousness time disappears completely and fundamentally changes. Eternity then becomes a moment, and a moment turns into eternity. In such a state we never have to worry about being bored.

Ultimately, God has no other purpose, nor any other mission, than to simply enjoy himself and be forever blissful. This is his natural state of being. He enjoys himself through his energies and his creation and by expanding into an infinite amount of independent living beings with whom he has a loving relationship. Therefore, he is not alone and not lonely, but he indeed creates a world and a universe, filled with living beings that can share and participate in his divine pastimes and stories. These living beings are part of his divine energies and emanations. The Vedas explain that these energies can be divided into three categories, which are:

1. His impersonal energy, *Brahman*, which is the substratum of both the spiritual and material creation. *Brahman* is the substance, the material, from which creation is constructed.

2. His personal energy, *jiva tattva*, consisting of living beings, conscious persons like ourselves. The living beings or *jivas* are partial copies of God himself, or as the Bible states: created in his image. The jiva's are qualitatively equal, but quantitatively different from God.

3. His plenary expansions, which are complete copies of himself that are both qualitatively and quantitatively equal to him. Since God is infinite he can expand and duplicate himself infinitely without ever being reduced or diminished. God, therefore, always remains simultaneously whole and complete, as the one God (so as not to be confused with polytheism) and remains the original Supreme Person. The explanation as to why God has plenary expansions will be discussed in the next chapter, but in summary it is to be able to have a direct and personal relationship with each and every living being.

The combination of these three energies represents the totality of his creation, although there is an important side note with reference to the use of the word 'creation'. As we have established previously, both God as well as his energies are eternal. Therefore, the use of the word creation can be misleading, since it implies that there is a moment of creation in time. This is not the case; the word creation should be interpreted as an emanation and transformation in which the emanation depends on the source, the creator. However, both the source and the emanation are eternal, which is contrary to Christian theology. Christian theology adheres to the principle of the so-called *creatio ex nihilo*, which means that God, the creator, creates his creation out of nothing. The Vedas reject the principle of *creatio ex nihilo* and they define creation as a transformation and emanation.

The purpose of creation, and the expansion and emanation of the energies of God, is for these expansions to participate and serve God in his pastimes and activities for his entertainment and pleasure. The living beings as part of creation enjoy themselves too, unlimitedly and fully as part of God's pastimes and activities, but in a supportive, subordinate and serving role. This is how the divine cosmic order has been arranged and how our roles in this cosmic order have been casted. These roles are entirely logical and necessary for a creation to be able to exist and to function at all. Only illusion can change this reality, or rather create the semblance of a changed reality. As we have explained previously, this material world has been created as an alternative, illusory reality for those living beings who, out of their free will, have the desire to be the controller and principal enjoyer, in other words to take on God's role in different degrees. We felt some desire to have what God has or be who God is. God facilitates these desires, but it inevitably leads to problems. In pursuing our own selfish desires and happiness beyond what we are entitled to, we will inevitably cause harm to other living beings, whether by accident or intention. In so doing we will interfere with their happiness, thereby disrupt the harmony of the universe and invoke the laws of karma.

Chapter 14

OUR DESTINATION: VAIKUNTHA

"If we find ourselves in the position that nothing in this world can satisfy us, then the most likely explanation is that we were made for a different world."

C.S. Lewis, famous British philosopher and writer

"The purpose of our journey through the material universe is the realisation that it does not have a purpose other than to make us realise that we should not be here."

Essence of the Vedanta philosophy

The four conclusions

The conclusion of this book can be summarized in four points. These four points mirror the essential principles of the Vedanta doctrine, the Bhagavad-Gita, the four noble truths of Buddhism, but also that of the Western religions Judaism, Christianity and Islam. This is not a coincidence. The scientific conclusions as presented in this book agree with the fundamental doctrines of the major world religions. Despite the major differences in culture, perception and ideology, the major world religions are in agreement on a number of essential principles. Even if quite a few followers of these various religions see this differently influenced by fundamentalism and sectarianism. Objectively, there clearly is a universal common denominator and the four propositions below reflect these.

It is important, especially in this period of globalisation, global communication and information networks, to emphasise the common ground and universal aspects of the various

In previous page, Gautama Buddha lived from 563 BC up to 483 BC in India and is the founder of Buddhism. Even though Buddha rejected the Vedic doctrine, his teachings were founded on and derived from the yoga systems and the Vedanta doctrine in many ways. Buddha rejected the Vedic doctrine for societal reasons, such as the corrupt caste system, abuse of animal sacrifices, and Brahmanic privileges. In many ways he was a spiritual and social reformer. Buddha mainly taught detachment from the material world, as is stated in his four Noble truths: 1. Existence in this world is characterised by suffering (dukkha); 2. There is a cause for this suffering (samudaya); 3. There is the possibility of end of suffering (nirodha); 4. There is a path that leads to the end of suffering (magga). By practising the eightfold path man can, in the end, be freed from suffering and he can reach the state of being enlightened, Nirvana. The eightfold path consists of: proper insight, proper intent, proper speech, proper action, proper earning of livelihood, proper effort, proper attention and proper concentration. This path is not really different from the various yoga systems such as jnana-, karma-, astanga- and bhakti-yoga. The Vedic texts describe Buddha as the 9th incarnation or avatar of Vishnu.

religious traditions. God is universal and he is not the possession of any religion or prophet. Religion exists to serve man and God and not the other way around. It is crucial that we should be able to lift religion up out of the domain of just faith and often times superstition and to place it into the domain of science and philosophy. Philosophical and theological differences, after all, can be discussed, debated and argued, which will lead to deeper insights and sharpening of the intellect. However, fundamentalism, sectarianism and fanaticism prohibit reasonable debate and exchange of information based on logic, science and philosophy. Fundamentalism and sectarianism are mostly instruments of the ego and oppose the development of spiritual consciousness. The key characteristics of authentic spiritual development are humility, service mindedness and a sincere desire to acquire truth in conjunction with a tolerance for other, opposing viewpoints. Unfortunately some adherents, across the spectrum of the various religious traditions, get caught up in a sense of being privileged, belonging to the chosen people. In a subtle, and sometimes not so subtle way, they develop a sense of superiority towards those whom they regard as less fortunate. Such a sense of superiority undermines every serious attempt towards true spiritual development, and despite noble aspirations, here too the ego can be the major spoiler.

Summarized: the four conclusions

These are the four points that sum up the conclusions of this book:

1. **God exists and the material universe is part of his creation. God is one, but simultaneously manifests himself as many because of the diversity of his emanations and energies. As living beings we are an integral part of God and one of his spiritual energies, possessing a conscious, personal identity that is eternal in time yet finite in space. We possess the same qualitative attributes as God, just in a limited amount. His existence has an all-determining**

effect on our vision on reality and on all aspects of our existence.

2. Our existence in this world has a clear purpose and it has a clear cause. Our lives are not a coincidence and they are not the result of an evolutionary accident. Above all, our existence in this world is characterized by suffering that we, as eternal souls, may experience life time after life time. This suffering is caused by a deep rooted desire for power and attention for the ego and therefore we tend to enhance and expand our own influence at the expense of others. The suffering we thus cause to others is something we have to experience ourselves based on the law of karma.

3. If we want to—and this choice is ours—we can find a path that leads us out of the labyrinth of this material world. This road is not easy, but certainly not impossible. Throughout the centuries the major world religions, in their own specific way and in accordance with time, place, and circumstances, have given guidelines about how we can liberate ourselves from the bondage of the material universe. The Vedic tradition uses the various yoga systems as methods for the development and expansion of consciousness in order to achieve the ultimate liberation from the material world.

4. There is another spiritual universe where the limitations and imperfections of material existence are not present. In the Vedas this universe is called Vaikuntha, which literarily means 'freedom of anxiety'. Even though the material universe is infinite from a human perspective, it is nevertheless only a miniscule part of and a domain of the mega-infinite spiritual universe. The material creation is described in the Vedas as a small cloud in a corner of the spiritual firmament. The purpose of our lives is to return to the spiritual universe, Vaikuntha, and to leave the material world behind us forever.

Vaikuntha, universe without anxiety and grief

Vaikuntha, the spiritual universe, is therefore our final destination, but it is also the place where we originated from. It has been stated many times before that the qualities and the attributes of the spiritual world are very difficult to describe and to comprehend. Even the attributes of the material universe, described in the Vedas as an inferior energy, are almost impossible to comprehend, especially if we analyse and investigate matter and the material energies at their deepest levels. We can therefore only imagine how difficult it will be to try to understand the spiritual universe, a universe that is constructed from superior, spiritual energy. At this effort we run out of hyperbolic superlatives, and our psychological, intellectual, and emotional limitations are an enormous hindrance to achieve such understanding. Yet, despite these massive limitations we are still able to develop some understanding, some level of visualization and some form of a mental image of the spiritual world. This is so because the material world is a shadow-like, upside down and perverted reflection of the spiritual world. In the Bhagavad-gita which contains the essence of Vedic philosophy a reference is made to a Banyan tree, which has as its unique feature that it has its roots upward and its branches downward, the opposite of a normal tree (Chapter 15 verses 1, 2, 3):

The Blessed Lord said: There is a banyan tree which has its roots upward and its branches down and whose leaves are the Vedic hymns. One who knows this tree is the knower of the Vedas. The branches of this tree extend downward and upward, nourished by the three modes of material nature. The twigs are the objects of the senses. This tree also has roots going down, and these are bound to the fruitive actions of human society. The real form of this tree cannot be perceived in this world. No one can understand where it ends, where it begins, or where its foundation is. But with determination one must cut down this strongly rooted tree with the weapon of detachment. Thereafter, one must seek that place from which, having gone, one never returns, and there surrender to that Supreme Personality of Godhead from whom everything began and from whom everything has extended since time immemorial.

The banyan tree is thereby compared to a tree standing at the bank of a river and whose image is reflected and mirrored in the water. The mirror reflection is obviously upside down where the

roots look like the branches and the branches look like the roots. A mirror reflection shows clear similarities to the original, and for this reason we are fundamentally able to understand the original, based on the reflection. By studying the reflection we can develop an understanding and a mental image of the original and this is what we effectively do when conducting scientific investigations into the world of matter. The scientific insights we have gained concerning the structure of matter and energy, the knowledge we have developed at a cosmological level, accompanied by refined and sophisticated philosophical deliberations provide us with an insight into God and his spiritual energies. However, there are two crucial differences between the reflection and the original: the reflection is made of a very different substance and the reflection is upside down with respect to the original. First of all the substance that makes up the material universe, matter, is of an inferior nature, even though it originates from and is a transformation of spiritual energy. In Sanskrit the material energy is called apara prakriti, which literally means inferior energy'. This in contrast with the spiritual universe, which is made of para prakriti or superior energy. The inferior nature of matter has to do with the second point, the upside down nature of the reflection. In this context upside down means that the attributes and the properties of matter are opposite to those of spirit. The attributes become perverted so that in its most extreme expression, good becomes bad, beautiful becomes ugly, truth becomes a falsehood and falsehood becomes truth, love turns into hate, and power turns into the abuse of power. In the end the properties that are displayed by either the material or spiritual energies, are determined and shaped by consciousness, the collective consciousness and collective karma of individual souls. In turn the supreme consciousness facilitates various properties to emerge in order to accommodate the individual and collective karmas. The qualification superior and inferior has specifically to do with the interaction of conscious individuals with their environment, and how living beings experience these interactions. The material energy in this respect is ruled by three modalities, or guna's in Sanskrit, which define the quality of the conscious experience. These modalities are Goodness or Sattva, Passion or Rajas and Ignorance or Tamas, and they rule the material energy in endless configurations and combinations. While the quality of Sattva closely resembles the superior qualities of the spiritual energy, when mixed in with passion and Ignorance these qualities get degraded and perverted. When Ignorance becomes the predominant driving force the degradation becomes complete showing the ugliest face of the material energy. The superior spiritual energy is also called Suddha Sattva or pure goodness, where passion and ignorance or completely absent. Suddha Sattva rules the spiritual world and entails the ultimate and perfect existence in a sublimely perfect, infinite universe, as described in some following paragraphs.

In a previous chapter the analogy was made between digital and analogue technologies in the world of computing, data processing and transmission. The atomic structure of matter was thereby compared to digital processing of information, which is highly efficient in the digital economy. From a spiritual perspective however the situation is the reverse. The atomic packetized structure of matter, requires constant work and attention in order to function. While atoms provide a basic functional structure, any attempt to refine these structures or control them and turn them to our advantage, requires a lot of effort, work and knowledge, as can be seen from the efforts required to conduct scientific research and create and develop technology. It took literally 400 years since the beginning of the scientific revolution to arrive at the point whereby the microchip could effectively be created and refined to today's levels of incredible computing power. Matter is contumacious and stubborn, and requires massive efforts to be moulded in order to produce a favourable outcome. Spiritual energy does not require such efforts, since everything is commanded by consciousness, instantaneously by merely wishing and thinking it. Considering the vast differences between matter and spirit, both in quality and substance, the study of the material universe can only provide us with a limited and indirect insight into God and his spiritual energies. In the end, just like in this world, direct empirical observation is absolutely necessary to gain real knowledge. The problem, of course, is that we do not have access to a direct, empirical observation of the spiritual world (i.e. most of us do not have this access). But, if we can acquire this information from someone who does have empirical access to this spiritual reality, then that is at least second best. For, after all, how many of us have access to the most advanced scientific research centre in the world, the particle accelerator in CERN Switzerland, to witness the recent discovery of the Higgs particle? Only a few privileged and qualified scientists that are part of the inner circle of the scientific community. We were able though to witness the report of this spectacular event on television and because of the authority of the scientists and the position of authority that we have bestowed upon them, we accept the authenticity of the account.

In a similar way there have been mystics, enlightened souls and divine incarnations that have had direct experiences of the spiritual world, or that have directly descended from the spiritual world. Even though their circumstances were quite different from the elite scientists of CERN, the knowledge procedure as such—namely the empirical observation of a reality that is inherently difficult to observe—is identical. Just as the report of the CERN scientists was reported by the media, the experiences of these mystics were described in the Vedic literature. As such, the Vedas—and especially the Puranas which are a part of the Vedic literature—give a detailed description of Vaikuntha. These descriptions, of course, appear to be like fairy tales. This is unavoidable; from our perspective in this material world and from this planet earth, the spiritual world is one big fairy tale. However, even a fairy tale does not even remotely describe the splendid reality of the spiritual world; it is in many ways indescribable. Advanced computer simulations can sometimes permeate this world of the fantastic and the unimaginable. There are Hollywood productions in which extra-terrestrial worlds were presented, both the heavenly and the hellish variations. Some of them were very impressive, and may have even provided a glimpse of super-natural and other worldly realities. Below a short synopsis is provided of the various descriptions of the

Beautiful symbolic illustration based on Vedic texts depicting the banyan tree with its complicated up-side down structure in which living beings become entangled during their journey through the material universe. The material universe is ruled by desire, expressed in four main areas of activities, 1. Morality, discipline and work, 2. Economic development, 3. Sense enjoyment, 4. Desire for Liberation. By development of good actions and good karma, a living being lays the foundation for an attempt at liberation from the material world. When good actions transform into devotional and spiritual activities based on and guided by the principles of Bhakti Yoga, spiritual consciousness will be awakened in due course of time enabling a living being to begin his journey back to Vaikuntha.

spiritual world as explained in the Vedas and the Puranas. We will have to try to understand and visualize these descriptions without the help of fantastic computer simulations and virtual reality, but we rather have to rely on our own imagination.

The nature of Vaikuntha

As was stated above, the material world is a mirror reflection of the spiritual world. Therefore, the spiritual world resembles the material universe in terms of structure and organization. There are planets, planetary systems and galaxies that are situated in the spiritual sky, which is brilliant, glowing and luminous. This in contrast with the material universe where space is hostile, void-like, freezing cold, and pitch dark. The only light in the material universe is the light that is generated by and emanates from stars, like our sun. On Pluto, for example, some billions of kilometres from the sun, everything is indeed freezing cold and

Our material universe is basically cold, dark, hostile and void-ish. From earth at a distance of 150 million kilometres from the sun, the sun looks bright and during the day lights up the entire sky. How different is the view from Pluto at an average distance of 5 billion kilometres from the sun. From this distance the sun appears just slightly bigger and brighter than other celestial bodies, and it has very little power. During Pluto's winter temperatures drop to -233 C, but don't bother waiting for the summer when temperatures reach –223 C.

pitch dark, where the sun appears as a tiny distant spot of light just slightly bigger than the other stars in a predominantly pitch black sky.

Contrary to the material world, the spiritual energy and everything that is made of it is pleasantly illuminating and brilliant. The spiritual universe is infinite and gigantic! The spiritual planets are also of an unimaginable size. The majestic abundance makes every earthly comparison hopelessly inadequate. In the Vedas a description is given of palaces made of pure precious stones, streets paved in gold, but also forests and landscapes of an unprecedented splendour and beauty, all made of the superior spiritual energy. Below you find a text from one of the Tantras, which are a part of the Vedic literature. This description refers to *Krishna-loka*, according to the Puranas the highest planet in the spiritual world. A part of *Krishna-loka* is *Vrindavan-dhama*, a place that takes up a special position in the spiritual world:

"Vrindavana-dhama is a place of ever-increasing joy. Flowers and fruits of all seasons grow there, and that transcendental land is full of the sweet sound of various birds. All directions resound with the humming of bumblebees, and it is served with cool breezes and the waters of the Yamuna River. Vrindavana is decorated with wish-fulfilling trees wound with creepers and beautiful flowers. Its divine beauty is ornamented with the pollen of red, blue, and white lotuses. The ground is made of jewels whose dazzling glory is equal to a myriad of suns rising in the sky at one time. On that ground is a garden of desire trees, which always shower divine love. In that garden is a jeweled temple whose pinnacle is made of rubies. It is decorated with various jewels, so it remains brilliantly effulgent through all seasons of the year. The temple is beautified with bright-colored canopies, glittering with various gems, and endowed with ruby-decorated coverings and jeweled gateways and arches. Its splendor is equal to millions of suns, and it is eternally free from the six waves of material miseries. In that temple there is a great golden throne inlaid with many

> *jewels. In this way one should meditate on the divine realm of the Supreme Lord, Sri Vrindavana-dhama." (Gautamiya Tantra 4)*

Below there is another description that indicates how the inhabitants of the spiritual world freely move about in the spiritual space in their own planes:

> *"The inhabitants of Vaikuntha travel in their airplanes made of lapis lazuli, emerald, and gold. The ladies in the Vaikuntha planets are as beautiful as the goddess of fortune herself. Such transcendentally beautiful ladies, their hands playing with lotuses and their leg bangles tinkling, are sometimes seen sweeping the marble walls, which are bedecked at intervals with golden borders, in order to receive the grace of the Supreme Personality of Godhead."*

The planets in the spiritual world are inhabited by countless living beings that all lead their lives in perfect happiness, without limitations of space and time. Time does not exist, all living beings are eternal, and there is only the eternal moment. Birth and death do not exist, nor do illness and old age. There is not any form of suffering, pain or fear and the spiritual bodies of the beings in the spiritual world are made of pure bliss and consciousness. The living beings there can take on any shape that they desire and they can travel without any limitation and without any effort to any planet or location in the spiritual world as they wish. The interaction between the beings in the spiritual world is harmonious, devotional, and loving. Despite all unprecedented opulence, splendour and exquisite beauty, the atmosphere is entirely dominated by love, service and devotion. The ultimate object of our love and devotion is God himself, who is personally present at each planet by means of his plenary expansions. God is the 'star' of the spiritual universe, just like we know stars in the shape of movie stars and rock stars in this world. Even our earthly stars, such as Elvis, the Beatles, Marilyn Monroe, and George Clooney enjoy a lot of fame and adoration and people cannot stop talking about these celebrities. Celebrities are the ultimate idols, desired by many, imitated by many, and venerated and worshiped by many. The human tendency to idolize and worship is normal and familiar, and it is a reflection of our original spiritual position, and reflects our natural position in our relation to God. All attributes and charms that we find attractive in a celebrity are attributes that God possesses infinitely and unlimitedly. Therefore, it is logical that God is the ultimate star of the spiritual world, amongst all the other stars. The beings in the spiritual world of course also love each other and they feel attracted to each other's exquisite features and attributes.

This is a society that we can only dream of: loving, respectful, joyful, dynamic, devoid of any negative emotions, and absolutely not boring. Selfishness, anger, fear and hatred are totally absent in the spiritual world, and therefore there are no Hitlers, Stalins or Maos. Sometimes it is argued that it is not possible to experience happiness unless we also experience pain and suffering. Happiness is seen as a contrast to suffering, and in many ways as the absence or relief from suffering. Therefore, so it is argued, the two, happiness and suffering, are intimately connected and intertwined, and we are not able to experience one without also experiencing the other. This argument is incorrect, and relates to the issue of emotional limitation, as discussed earlier. Light is not the absence of darkness, rather darkness is the absence of light. In a similar way, happiness is not the absence of suffering, rather suffering is the absence of happiness. Light is the positive presence of something, whereas darkness is the negative absence of something, i.e. light, the two are not equal, and to pose them as such is a false equivalence. Happiness is the positive presence of an emotional state, fed by the presence of expanded consciousness, whereas suffering is also an emotional state and a real experience, but fed by the absence of expanded consciousness and the absence of happiness. So therefore as has been argued in the previous chapter, perpetual happiness is certainly possible, without having to rely on the presence of suffering. In fact according to Vedic doctrine, the natural state of every living being is a state of perpetual happiness, and that's why every living being, deep inside, is always longing, consciously or unconsciously, to re-obtain that state.

There is no work to be done in the spiritual universe, for everything is already existing in a perfect state. Every wish, every desire is fulfilled; just by desiring it; the desire instantly becomes reality. If there is no work to be done, then there is just entertainment. Our earthly existence is roughly divided into three activities, namely work, relaxation and recreation, and sleep. In the spiritual world there is no work and since spiritual bodies do not become tired, entertainment and recreation is the main activity of every living entity.

People sometimes get into a state of panic when they are not able to work anymore and have to spend their time focused on their hobbies and other leisure activities. This was discussed in the previous chapter, in reference to boredom and monotony. Boredom is caused by emotional limitation, and when the emotional limitation is removed, boredom and any other negative emotions will disappear, just like darkness disappears automatically when there is light. The idea of angels on a cloud playing harp for eternity is a caricature in every possible way. First of all, the spiritual world is infinitely more than angels, harps, and clouds as we can see from the aforementioned Vedic

On the next page a schematic and colourful depiction of the spiritual and material universes according to the Vedic texts. The image depicts the two distinct realities, the spiritual universe divided into many sub-domains, and the material universe. According to Vedic cosmology the material universe is actually a multiverse, but not the type of multiverse as described in chapter 2 of this book. In the Vedic multiverse all universes are actually part of the same domain, the same realm, which is the realm of the material energy, called apara-prakriti, which is basically ruled by the same material laws of nature. The spiritual and material universe are separated by an infinite ocean of spiritual and luminous energy, called the Brahmajyoti, and made of pure spiritual energy called Brahman. Beyond the Brahhmajyoti lies Vaikuntha, the magnificent spiritual world, infinite and unlimited, and made of Sat, Chit Ananda, eternal existence, full knowledge and unlimited bliss. Every living being is spiritual and originated from the spiritual realm, even though temporarily captured by the material energy. And every living being is destined to ultimately complete his journey through this temporary realm of birth and death in order to return to Vaikuntha, the eternal spiritual world.

descriptions of the spiritual world. What is even more important is that the living beings in the spiritual world are emotionally fulfilled; they are in a perpetual state of bliss. The soul, in its liberated state, has gained access to the infinite reservoir of emotions of all varieties and of such an intensity that it is literally bathing in them. The core of these emotions is love, the love for God, for other beings and love for the creations of God. For this reason the mere empirical observation of the spiritual world would not be enough to understand the spiritual world. The real understanding, which will give depth to the empirical observation, is the emotional fulfilment of the soul, which will make the experience of the spiritual world complete. In other words, even if we are able to create a brilliant computer simulation that can represent the spiritual world in all its splendour and riches and glory, without the depth of spiritual emotions of love, delight and bliss this experience will be insufficient. If we want to experience the spiritual world and if we want to return to this world, we will have to take the right steps to achieve this goal. Those steps and the accompanying efforts will not be simple and they will not be easy, but they are certainly not impossible. With the right and proper effort, the right knowledge and the right guidance driven by a sincere desire, we can all achieve this desirable goal, which is according to Vedanta doctrine our eternal birth right and ultimate destination.

List of Sources and References

'1   *Quotation of Einstein, cited by Fred Hoyle in 'The Intelligent Universe*

2   *Richard Dawkins*

3   *Bertrand Russell "Religion and Science (Oxford University Press, 1961)*

4   *Richard Dawkins in the "God delusion" page 35*

5   *Richard Dawkins in the "God delusion" page 35*

6   *Robert J. Spitzer in 'New proofs for the existence of God*

7   *Chandra Wickramasinghe in Fred Hoyle and Chandra Wickramasinghe 'Evolution from Space' (Simon & Schuster, New York 1984)*

8   *Fred Hoyle in Fred Hoyle and Chandra Wickramasinghe 'Evolution from Space' (Simon & Schuster, New York 1984)*

9   *Fred Hoyle, 1981, cited by Robert J. Spitzer, cited in his book 'New Proofs for the Existence of God', page 73. Original quotation by Fred Hoyle: 'The Universe: Past and Present*

Reflections' in Engineering and Science, (November 1981)

10   *Martin Rees 'Just six numbers, page 166*

11   *Richard Dawkins in 'Climbing Mount Improbable' (W.W. Norton, New York 1996)*

12   *Patrick Glynn 'God, the evidence', page 31. Here he cites Fred Hoyle in 'The Origin of the Universe and the Origin of religion'*

13   *Leonard Susskind in 'The Cosmic Landscape' page 129*

14   *Robert J. Spitzer in 'New proofs for the existence of God'*

15   *National Geographic documentary on the 'origin of human species', Sunday 28 May, 2006. If you take the DNA of a cell, and lay it out it would stretch 3000 times the distance (back and forth) from the earth to the moon which is 400.000 km, which is 2.4 billion km (3.000 x 800.000)*

16   *William A. Dembski 'Intelligent Design' page 165, 166*

17   *Harun Yahya in 'Darwinism refuted', page 197, where he cites Robert Shapiro 'Origins: A Skeptic's Guide to the Creation of Life on Earth' (Summit Books, NY 1986)*

18   *Chandra Wickramasinghe in Fred Hoyle and Chandra Wickramasinghe 'Evolution from Space' (Simon & Schuster, New York 1984)*

19   *Fred Hoyle in Fred Hoyle and Chandra Wickramasinghe 'Evolution from Space' Simon & Schuster New York, 1984*

20   *Stephen C. Meyer, 'Signature in the Cell'*

21   *Richard Thompson 'Mechanistic and nonmechanistic science' page 122*

22   *Richard Dawkins in 'Climbing Mount Improbable' (W.W. Norton, New York 1996)*

23   *Robert J. Spitzer 'New proofs for the Existence of God'*

24   *Robert J. Spitzer 'New proofs for the Existence of God'*

25   *Martin Rees 'Just six numbers'*

26   *Robert J. Spitzer 'New proofs for the Existence of God'*

27   *Jacques Monod (1972) 'Chance and Necessity'*

28   *David Hume (1779) 'Dialogues Concerning Natural Religion', pages 155-156. Quote from Michael Denton, 'Evolution: A theory in Crises' page 40*

29   *Marsaglia G., and Zaman, A. (1993) 'Monkey tests for random number generators', in Computers & Mathematics with Applications 26*

30   *Sir Fred Hoyle in 'The Intelligent Universe' page 25*

31   *Richard Dawkins in "The Blind Watchmaker" (1986) page 278*

32   *Mayr, E. (1972) 'The Nature of the Darwinian Revolution', page 988, quotation in Michael Denton's 'Evolution: a Theory in Crisis' page 357*

33   *William A. Dembski in 'Intelligent Design' page 18*

34   *Richard Thompson in 'Mechanistic and nonmechanistic science', remark made at a meeting of evolutionists in Chicago (probably in 1980)*

35   *John Ankerberg and John Weldon, in 'Darwin's Leap of Faith'*

36   *Robert L. Carroll, Patters and Processes of Vertebrate Evolution, Cambridge University Press, 1997*

37   *Charles Darwin, The Origin of Species: A Facsimile of the First Edition, Harvard University Press, 1964*

38   *Charles Darwin, The Origin of Species by Means of Natural Selection, The Modern Library, New York*

39   *Robert L. Carroll, Patters and Processes of Vertebrate Evolution, Cambridge University Press, 1997*

40   *Francis Hitching in 'The Neck of the Giraffe: Where Darwin Went Wrong', (Tichnor and Fields, New Haven 1982)*

41 *H.J. Muller in 'How Radiation Changes the Genetic Constitution' in Bulletin of Atomic Scientists, 11 (1955), p. 331*

42 *David A. Demick, 'The Blind Gunman', Impact, no. 308, February 1999*

43 *Michael Behe 'Darwin's Black Box'*

44 *Jonathan Wells 'Icons of Evolution"*

45 *Jonathan Wells 'Icons of Evolution'*

46 *Richard Thompson in 'Mechanistic and nonmechanistic science', page 195, where he cites from Darwin 'On the Origin of Species'*

47 *John Ankerberg and John Weldon in 'Darwin's Leap of Faith', page 125, where they cite from Charles Darwin 'The Origin of the Species' (Baltimore, MD: Penguin Books, 1974)*

48 *John Ankerberg and John Weldon in 'Darwin's Leap of Faith', where they cite A.E.Wilder-Smith in the publication 'The Natural Sciences Know Nothing of Evolution'*

49 *B.L. Woolley in 'Adherents of Permanent Revolution: a History of the Fourth (Trotskyist) International', University Press of America*

50 *John Ankerberg and John Weldon in 'Darwin's Leap of Faith'*

51 *Center for Science and Culture, review by David Klinghoffer about the movie 'Expelled, No Intelligence Allowed'*

52 *Leonard Peikoff in 'The Ominous Parallels'*

53 *Leonard Peikoff in 'The Ominous Parallels'*

54 *John Ankerberg and John Weldon in, 'Darwin's Leap of Faith'*

55 *Sir Fred Hoyle in 'The Intelligent Universe' page 25*

56 *Corey S. Powell in 'God in the equation' where he cites Einstein – page 3*

57 *Fritjof Capra, 'The Tao of Physics' page 66*

58 *Fritjof Capra, 'The Tao of Physics' page 73*

59 *Fritjof Capra, 'The Tao of Physics' page 51*

60 *Fritjof Capra, 'The Tao of Physics' page 51*

61 *Fritjof Capra in 'The Tao of Physics', page 50, where he cites Werner Heisenberg: W. Heisenberg op. Cit.*

62 *Source: Wikipedia http://nl.wikipedia.org/wiki/Schrödingers_kat*

63 *Ludwig Wittgenstein in Tractatus Logico-Philosophicus*

64 *Richard Dawkins, 'The God Delusion'*

65 *Martin Rees in 'Just six numbers' page no 45*

66 *Martin Rees in 'Just six numbers' page no 83*

67 *Martin Rees in 'Just six numbers' page no 92*

68 *Bill Bryson 'Een kleine geschiedenis van bijna alles'*

69 *Bill Bryson 'Een kleine geschiedenis van bijna alles'*

70 *Patrick Glynn in 'God, the evidence'. Here he cites Fred Hoyle in 'The Origin of the Universe and the Origin of Religion', Chapter one, page 28*

71 *Steven Weinberg in, 'The First Three Minutes' page 8*

72 *William A. Dembski in ‚Intelligent Design'*

73 *Richard Dawkins in 'The God Delusion', chapter 3, page 135*

74 *Richard Dawkins in 'The God Delusion', chapter 4, page 184*

75 *Paul Davies in 'The Goldilocks Enigma', who cites in turn Lucretius: "Ex nihilo, nihil fit" in 'On the Nature of the Universe' (De rerum natura), translated by R.E. Latham (New York: Penguin, 1951)*

76 *William A. Dembski in ‚Intelligent Design'*

77 *Richard Dawkins in 'The God Delusion'*

78 *John S. Feinberg in 'the many faces of evil, theological systems*

and the problems of Evil' cited on location 4492, referring to William Rowe, 'The problem of Evil and some varieties of Atheism'

79   *Bergson 1935*

80   *A.C. Bhaktivedanta Swami in 'Bhagavad-gita', chapter 2, verse 13*

81   *Stephen Hawking in 'A Brief History of Time'*

82   *Victor Hugo in 'Les Miserables' (1862, pt. 2, bk. 7, Ch. 6).*

83   *Stephen Hawking in 'A Brief History of Time', page 190*

84   *Ludwig Wittgenstein in Tractatus Logico-Philosophicus vers 6.44*

85   *Bernard Haisch in 'The God Theory'*

86   *Velmans & Schneider, pp.1-6 2007, reference in Wikipedia*

87   *Richard Dawkins*

88   *Robert J. Spitzer 'New proofs for the Existence of God', page 74 where he cites Bruce Gordon*

89   *A.S. Eddington, 'The Nature of the Physical World', de Gifford-lectures by Eddington at the University of Edinburgh from January up to March 1927, Kessinger Publishing, 2005*

90   *Fritjof Capra in 'The Tao of Physics'*

91   *Fritjof Capra in 'The Tao of Physics'*

INDEX